CT-5

CLASSROOM TRAINING HANDBOOK

NONDESTRUCTIVE TESTING

Eddy Current

Published by ***PH D****iversified*, Inc.
5040 Highway 49 South
Harrisburg, NC 28075

Printed in the United States of America
ISBN 1-886630-17-8

PREFACE

Classroom Training Handbook - Eddy Current Testing, CT-5 is one of a series of training handbooks designed for use as the primary Level I "text" when NDT is taught in a traditional classroom setting. The instructor would typically assign chapters and discuss the material in a classroom lecture format.

This **Classroom Training Handbook** is most effective when used in conjunction with the Programmed Instruction Handbooks, PI-5, Eddy Current Testing (2 volumes). The Programmed Instruction Handbooks present essentially the same entry-level material but in a **self-study format**. The instructor can make assignments in the classroom handbook and, as a supplement, the student can read corresponding information in the self-study handbooks.

The **Classroom Training and Programmed Instruction** Handbooks can both be used successfully as refresher material for Level II training provided that sufficient industry specific equipment, specifications and applications are added to the course outline.

Other **Classroom Training Handbooks** in the series include:

- CT-2 Liquid Penetrant Testing
- CT-3 Magnetic Particle Testing
- CT-4 Ultrasonic Testing
- CT-6 Radiographic Testing

It is recommended that PI-1, Introduction to Nondestructive Testing, be completed before starting this book in order to have the benefit of certain basic metallurgy information that will make this book easier to understand.

ACKNOWLEDGMENTS

Publishing and Printing

Revision Editor . . . . Dr. George Pherigo, ***PH D****iversified*, Inc.

Production Editor . . Ms. Mary Lou Hollifield, ***PH D****iversified*, Inc.

Proofreading . . . . Ms. Jean Pherigo, ***PH D****iversified*, Inc.

Technical Content Revision

Technical Editor . . . . Mr. Jim Cox, ZETEC, Inc.

Production Assistant . . . . Ms. Kathie Tobin, ZETEC, Inc.

This handbook was originally prepared by the Convair Division of General Dynamics Corporation under contract to NASA and was identified as N68-28792. This book is part of a series of books, commonly known as the General Dynamics Series, that has been the basis of many industrial NDT training programs for over 20 years.

Now, after several decades of widespread use, the entire series has undergone a major revision. The revised material no longer concentrates on applications in the aerospace industry, but instead, covers a wider range of industrial applications and discusses the newest techniques and applications.

Mr. Jim Cox has been the principal editor of the revised material in this text. Using his nondestructive testing experiences in several industries, including work at the EPRI NDE Center and Zetec, Inc. he has updated the text to better suit the entry-level NDT technician/engineer.

TABLE OF CONTENTS

CHAPTER 1

ELECTROMAGNETIC INDUCTION

Eddy current testing is based on the principles of *electromagnetic induction*. "Electromagnetic induction"—two very scientific-sounding words that are used to identify a principle that allows you to use electricity that has been generated hundreds of miles away, a principle upon which the actual generation of the electric current is based, a principle that causes your electric motor to operate, and now a principle upon which a broad field of nondestructive testing is based.

The word "electromagnetic" simply means that electricity and magnetism are used. "Induction" is a form of the word "induce" which means "to bring about" or "to cause." In fact, under certain circumstances the flow of electricity can cause magnetism, and under certain circumstances magnetism can cause the flow of electricity.

By the year 1820, scientists had discovered that when current from a battery was sent through a coil of wire a magnetic field was set up in and around the coil. The magnetic field was present only during the time the current flowed through the coil. They had discovered how to use electricity to make magnetism and they thought that somehow magnetism could be used to make electricity.

Some 12 years later, in 1832, a man named Faraday was experimenting with some coils of wire and a battery. He noticed that when he connected one coil to the battery he got an electrical current through a second coil, placed near the first coil, for just an instant. He also found that when he disconnected the battery he got an electrical current through the second coil for just an instant; but, he noticed the second current was in the opposite direction of the first current.

He knew that somehow the two coils were affecting each other. The first coil was *inducing* a current in the second coil, but only when he turned the battery on and off. He reasoned that the magnetic field could be the coupling between the two coils. But, since the currents occurred only at the moment when the battery was turned on and off, it could only be the *change* in the magnetic field that caused the current to flow in the second coil.

Here's what Faraday's experiment looked like.

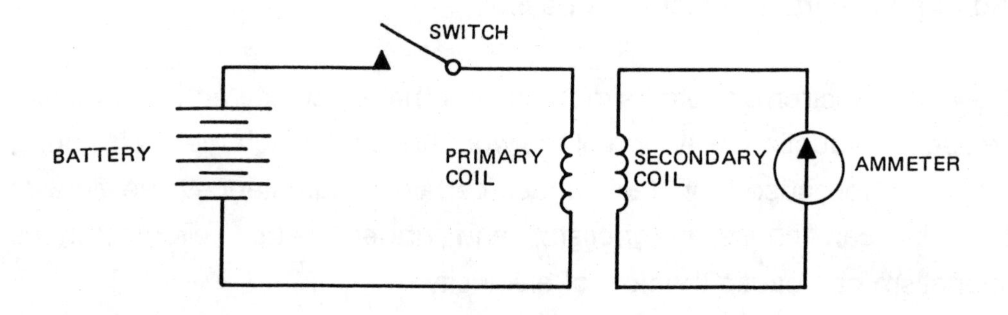

Figure 1-1
Faraday's Original Electromagnetic Induction Experiment

The next logical step was to make different changes in the setup and see what effect they had. For example:

- change the number of turns in the primary coil.
- change the physical size of the primary coil.

- change the amount of current in the primary coil.
- change the number of turns in the secondary coil.
- change the physical size of the secondary coil.
- change the spacing between the coils.

All of these things can be changed. They are called "variables." We won't, at this point, go into the effect that each of these variables had on the amount of current that was induced in the secondary coil. It is enough to say that each and all of these variables had an effect on the current induced in the secondary coil. It changed—in one way or another.

Now that we have established what a variable is, let's get on with electromagnetic induction. We have described how Faraday was able to produce an electrical current in a secondary coil by *changing* the magnetic field surrounding a primary coil. Faraday reasoned that the current was produced by the *change* in the magnetic field and not by the simple presence of the field. In other words, so long as the magnetic field around the primary winding did not vary (was held constant), no electrical current was induced in the secondary coil. Thus, utilizing the theory of a magnetic field, current was induced *only* when the lines of force of the magnetic field moved past the coil. Figure 1-2 illustrates this.

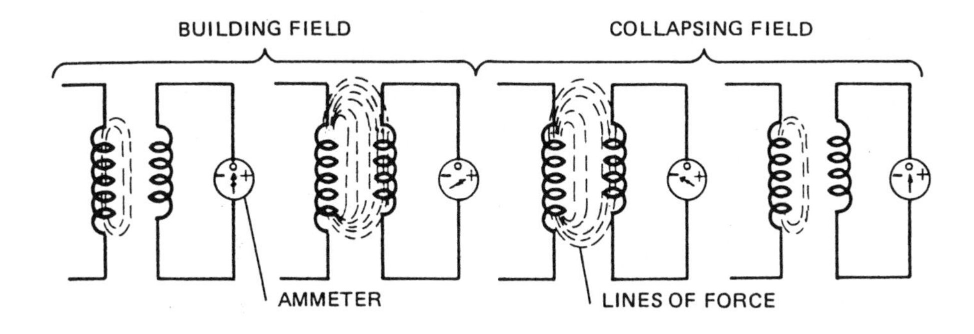

Figure 1-2
Time Varying Magnetic Field About a Coil

Now, if this were true, as it appeared to be, then it should be possible to induce a current by moving a coil through a magnetic field as shown in Figure 1-3. A coil of wire was placed in the open end of a horseshoe magnet and given a spin; electricity was induced in the coil. However, the current induced did not travel in the same direction through the coil at all times, nor was it a constant value. This situation required analysis.

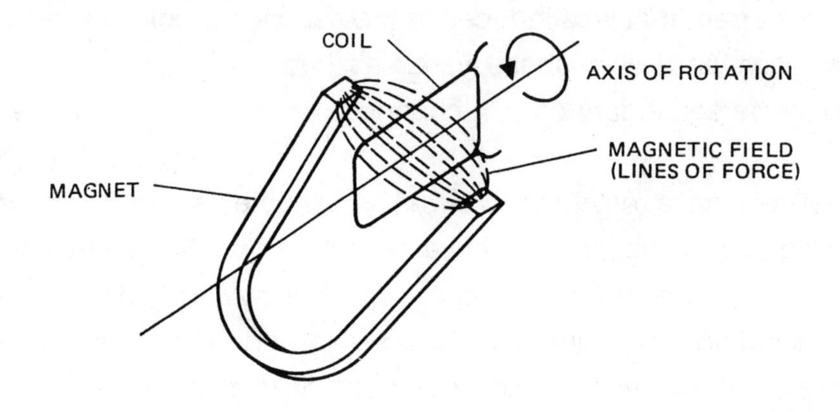

Figure 1-3
Moving a Coil Through a Magnetic Field

The analysis of the current being produced showed that the current started out at zero, rose to a maximum value in one direction, returned to zero, rose to a maximum value in the opposite direction, and returned to zero in one complete revolution of the coil. All of this could be explained by the theory of electromagnetic induction.

The induction of the electrical current into a coil is due to the relative motion between the magnetic field and the coil. It makes no difference whether the magnetic field is expanding and contracting past the coil or whether the coil is moving through the magnetic field. The *relative* motion is the same. Thus, a current is induced in the coil in either case. Figure 1-4 shows another model that might help explain this point.

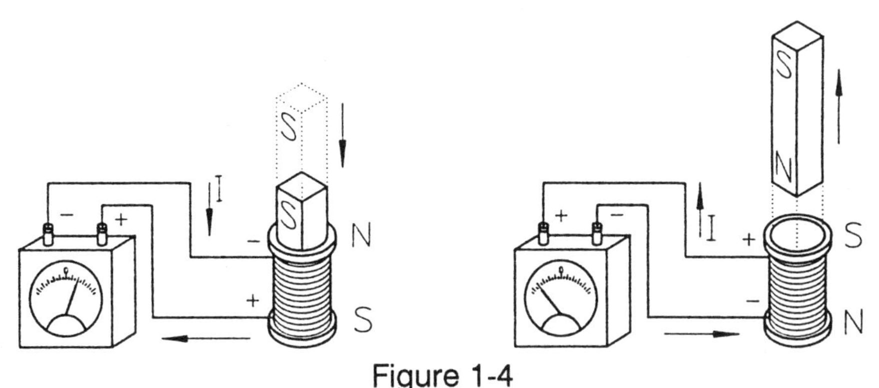

Figure 1-4
A Bar Magnet Moving Through a Coil of Wire

Let's go back and examine what might happen when a single wire is moved through a magnetic field. The wire shown in Figure 1-5 being moved downward through the magnetic field, has a current induced in it as it passes through the lines of force. Suppose that the lines of force are thought of as having a direction from the north pole to the south pole, and suppose the current induced in the wire travels away from you as you look at Figure 1-5.

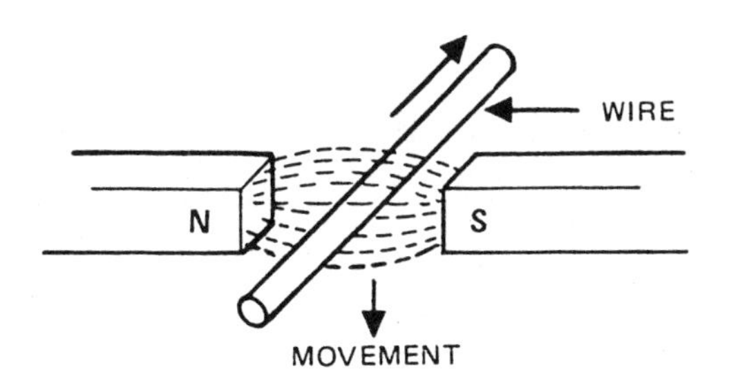

Figure 1-5
A Single Wire Moving Through a Magnetic Field

The direction of the current will change when the direction of the movement of the wire is changed. There is a rule for determining the direction of the current induced in the wire, but you will not have any need for it in eddy current testing, so we will not bother to learn it. It is sufficient that you know that there are three ways to cause the current to change its direction in the wire. First, you could change the direction of the magnetic field (difficult to do in a permanent magnet); second, you could change the direction that the wire is moving through the field; or third, you could swap ends with the wire (which is exactly what happens when a coil is rotated in a magnetic field).

Now, let's bend the wire into a "U" shape, insert it into the magnetic field, and rotate it around the axis as shown in Figure 1-6. The segment of wire A-B will be coming down through the field while segment C-D is coming up through the field.

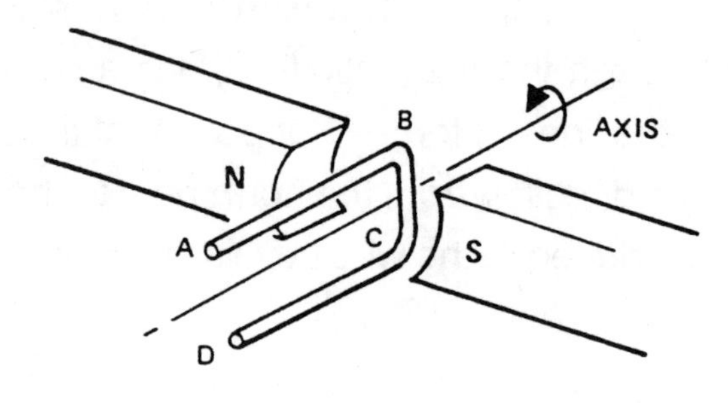

Figure 1-6
Rotating Coil Within a Magnetic Field

Figure 1-7 shows that the current through the segment C-D is added to the current through A-B so that we have current flow now from A to D (in that direction).

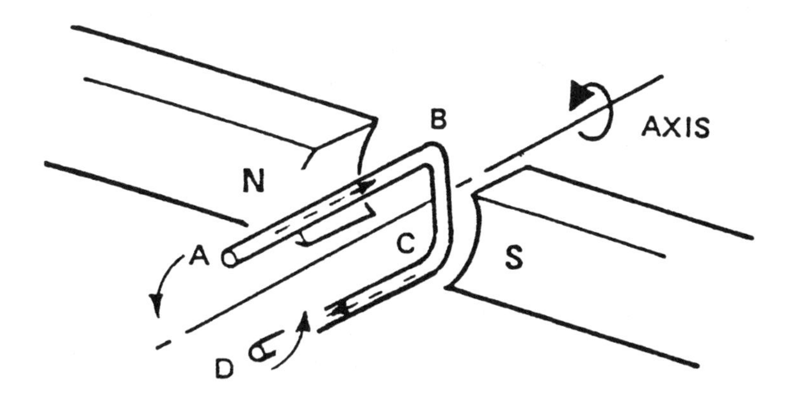

Figure 1-7
Rotating Coil Within a Magnetic Field - Position Unchanged

As illustrated in Figure 1-8, if we keep rotating the wire until the segment D-C is coming down through the field and segment A-B is moving up, then the current in segment D-C is flowing from D towards C and then current in segment A-B will be moving from B to A. We now have a current in the wire from D to A.

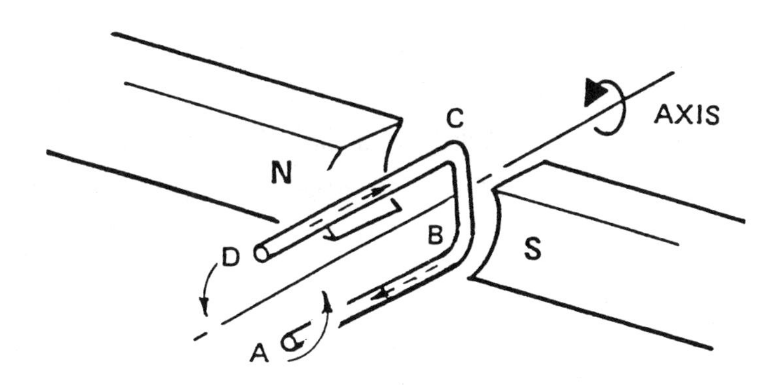

Figure 1-8
Rotating Coil Within a Magnetic Field - Position Rotated

Now let's compare the two situations shown in Figure 1-6 and 1-7. In the first instance the current flowed through the wire from A to D; in the second instance the current flowed through the wire from D to A. Imagine that the wire is wrapped to form several loops and then spun on its axis in the magnetic field as shown in Figure 1-9. The current through the loop will change its direction at every half-turn that the coil makes as it rotates.

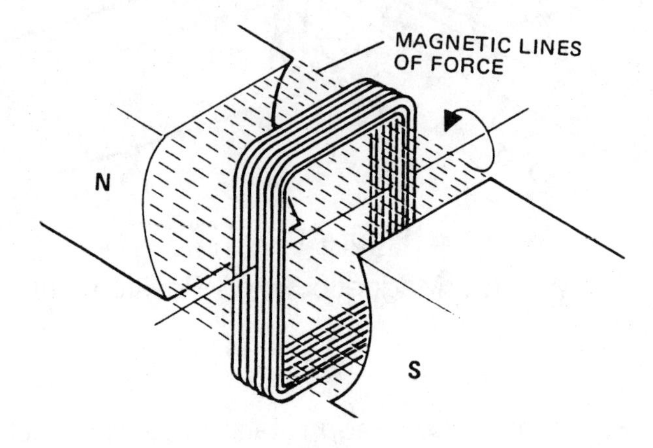

Figure 1-9
Rotating Coil Within a Magnetic Field
Coil Windings Parallel to Flux

Figure 1-10 shows a coil in a position where the *plane* of the coil is across the magnetic lines of force. Notice that as the coil moves, the top windings and the bottom windings are moving in a direction that is *parallel* to the lines of force. Since the direction of movement is parallel to the lines of force, *no lines of force are being crossed; therefore, no current is being induced* in the coil.

In Figure 1-10, the coil has rotated 90°. The windings now lie parallel to the lines of force, but the movement of the coil sides is perpendicular to the lines of force. At this point, as the coil turns, it is passing through (or crossing) the maximum number of magnetic lines of force.

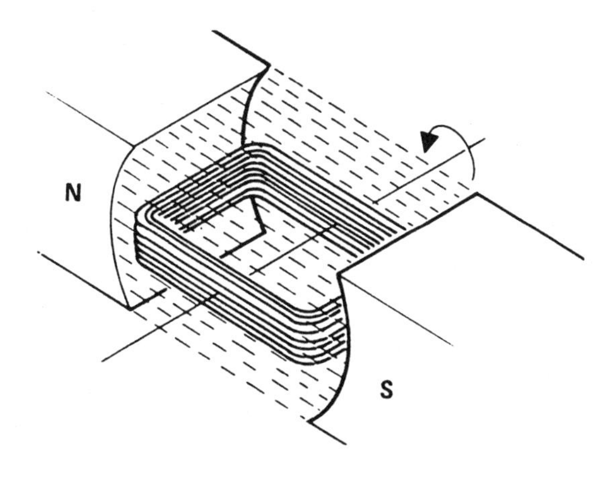

Figure 1-10
Rotating Coil Within a Magnetic Field -
Coil Windings Perpendicular to Flux

Current is induced in a coil only when the coil is cutting *across* the magnetic lines of force. By "cutting across" we mean that the *motion* of the coil is such that the wires in the coil pass through the magnetic field in some direction that is *not parallel* to the lines of force. The more lines that are being cut in a given period of time, the more current induced. Now, let's draw a graph so that we can visualize what is happening to the current as the coil rotates.

As shown in Figure 1-11, the instant the coil is at position A (0°) the current is zero; the instant the coil is at position B (90° of rotation) the current is maximum in one direction; at position C (180° of rotation) the current is zero; at position D (270° of rotation) the current is maximum in the opposite direction; and at position E (360° of rotation) the current is back at zero. The curve that results from this plotting of current values against coil position is called a *sine wave*.

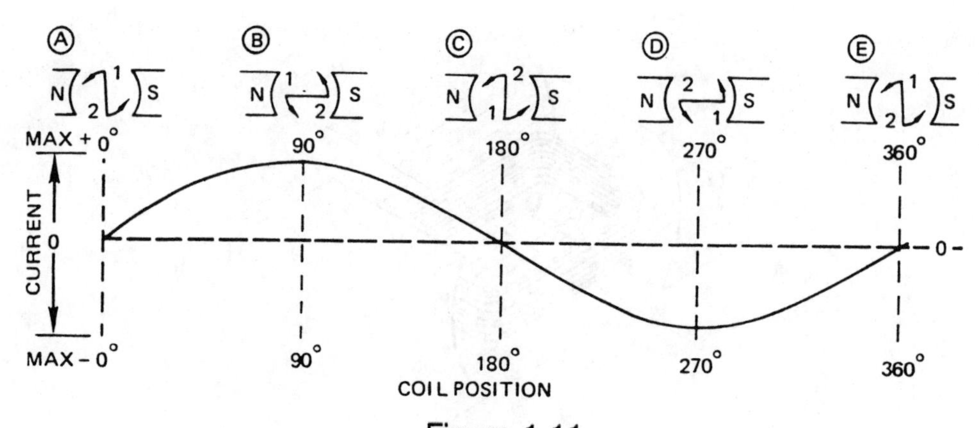

Figure 1-11
Current Output Based on Coil/Flux Relationship

Figure 1-12 below shows a concept of utmost importance. On our graph we have labeled the horizontal axis in degrees (0°, 90°, 180°, 270°, 360°) which refer to the position of the coil. These could just as well have been units of time. In fact, they *are* units of time—90° being a measurement of the amount of time it took the coil to travel from 0° to 90°.

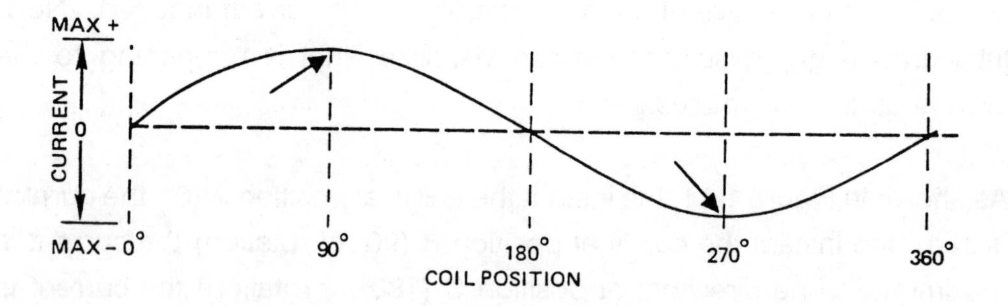

Figure 1-12
Current Output Based on Coil/Flux Relationship

As we progress, you will run into such statements as "the voltage lags behind the current in time by 15°." It is by virtue of the relationship between the rotating coil and time elapsed that we can measure *time in degrees*. Essentially, we have already measured time in degrees when we said that maximum positive current is obtained at 90°. It is easier to work in degrees of rotation when explaining *electromagnetic induction* than it is to work in units of time (seconds, milliseconds, etc.).

So far we have shown you the current generated as a coil makes one revolution in a magnetic field. What we have shown you is one *cycle* of current. An alternating current generator, of course, does not stop after one revolution. As the coil continues to rotate through additional revolutions, the generator puts out more cycles of current, each cycle corresponding to one revolution of the coil.

The number of current cycles put out by the generator in one second is called the *frequency* of the alternating current. The usual household alternating current, for example, has a frequency of 60 cycles per second. The term hertz (abbreviated Hz) is the term that is used instead of cycles per second. Household current, then, is at 60 Hz. In eddy current testing we often use frequencies in the kilohertz range (times one thousand cycles per second [kHz]) or sometimes even the megahertz range (times one million cycles per second [MHz]). These kinds of frequencies are not generated by rotating a coil through a magnetic field. Instead, special electronic circuits convert the 60 Hz frequency to the much higher frequencies used in eddy current testing.

Now that we know what happens when a coil is rotating in a magnetic field, let's go back and look at electromagnetic induction between two coils, but instead of using a battery, we will supply the primary coil with a source of alternating current as shown in Figure 1-13.

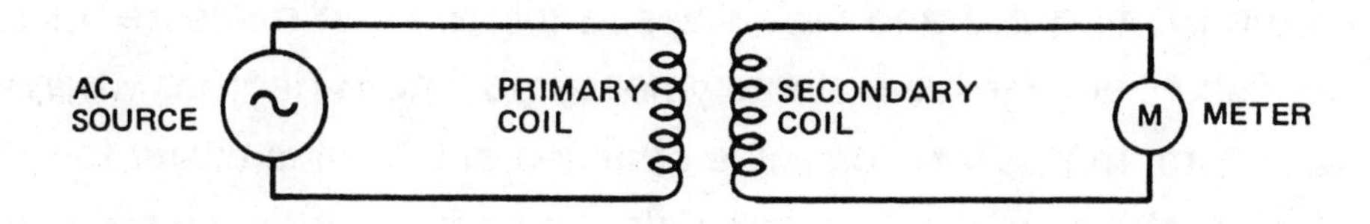

Figure 1-13
Electromagnetic Induction by an AC Source

The alternating current from the power source is in the form of the sine wave that was generated by rotating a coil in a magnetic field. The important point is that the current in the primary coil is constantly varying. It goes from zero to a positive maximum, back to zero, then to a negative maximum, and finally back to zero again. This is one full "cycle."

The magnetic field in the primary coil is varying in exactly the same manner as the current. We now have a situation where the magnetic field is building up in one direction, collapsing, building up in the opposite direction, collapsing, and so on (Figure 1-14). Since this field intercepts the secondary coil, a current is constantly being induced in the secondary coil because the lines of force are cutting across the wires forming the secondary coil.

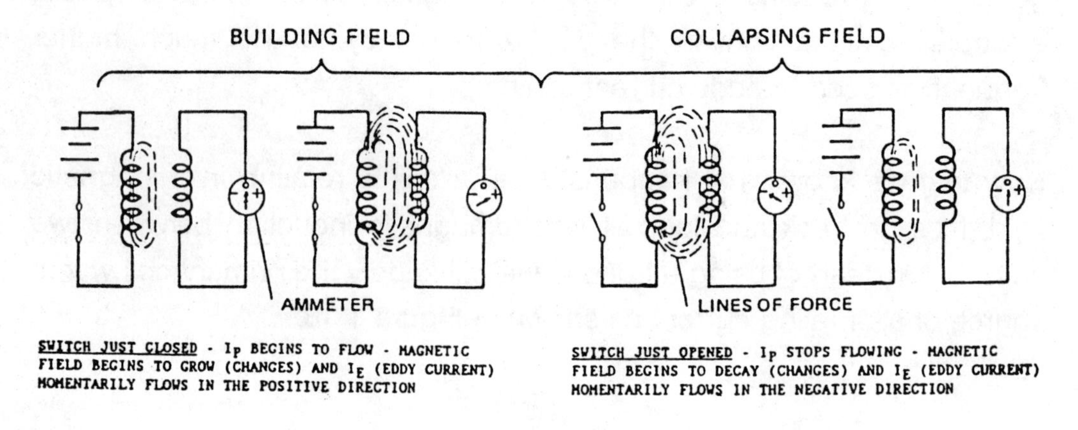

Figure 1-14
Time Varying Magnetic Field About a Coil

In order for the secondary coil to carry current, it must be made of an electrically-conductive material. For example, no current would be induced in a coil of cotton string, since cotton will not conduct electricity.

In general, metals are the best conductors of electricity, but there is a difference in conductivity even between metals. Silver has the best conductivity of all the metals, while titanium has the lowest conductivity. This means that silver has *less* resistance to the flow of electricity than titanium. The amount of current induced in the secondary coil is affected by the conductivity of the material in the secondary coil. A higher conductivity allows more current to be induced than a lower conductivity. This is an important point to remember in eddy current testing.

Now let's look at another point of extreme importance in eddy current testing. Let's see what occurs when an alternating current is applied to a coil. In Figure 1-15, we connect a voltmeter to measure the voltage across the coil and put an ammeter in the circuit to measure the current. If we plot the instantaneous readings of the instruments on a graph, we find that the voltage rises to a maximum before any current begins to flow. Then, while the voltage is decreasing to zero, the current is increasing to a maximum. From the chart you can see that the current through the coil lags *behind* the voltage by 90°.

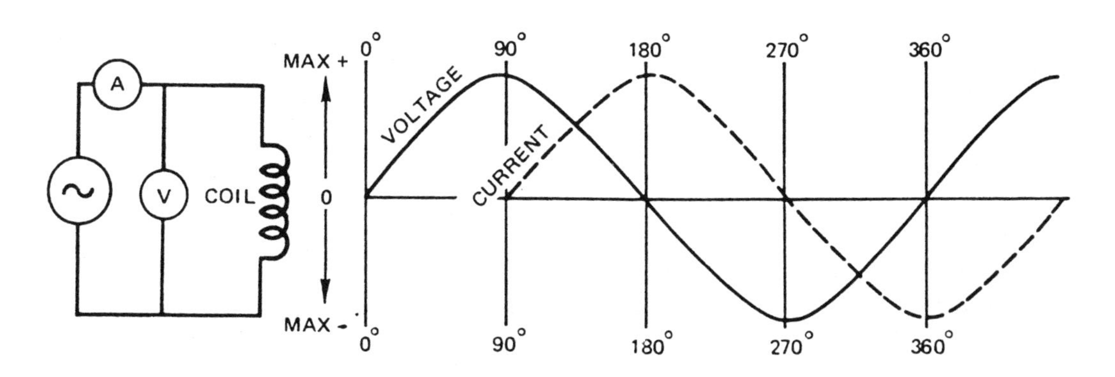

Figure 1-15
AC Coil Current and Voltage Characteristics

Figure 1-16 shows why this occurs. Let's consider for a moment a coil with one turn slightly separated from the rest of the coil. The alternating current through each turn produces a constantly varying magnetic field that cuts across the other turns in the coil, thereby inducing a current in each of the other turns of the coil. Each turn of the coil has multiple magnetic fields interacting with it and inducing current flow at slightly different strengths and directions. These self-induced currents *oppose* the original current in part of the cycle and aid the original current in another part of the cycle. The net effect is that the resultant current is shifted out of phase with the voltage. (It is delayed in time.)

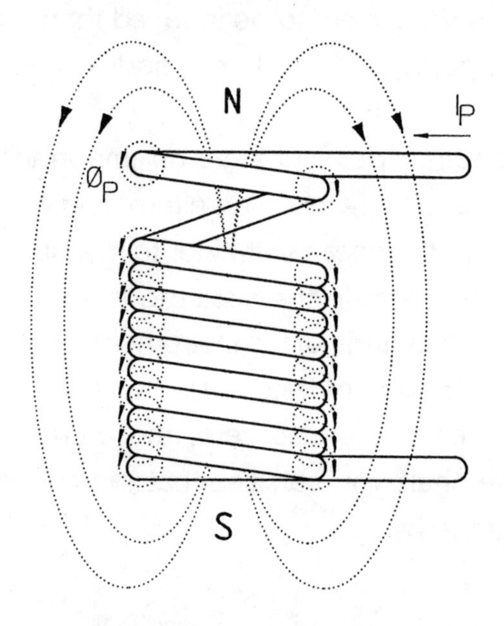

Figure 1-16
Magnetic Field About a Coil

Every turn in the coil induces the same effect in every other turn. The overall effect is that the current through the coil lags behind the voltage by 90°. This effect that causes the current to lag behind the voltage is called *inductive reactance* (X_L).

Now let's take a look at another factor
through a circuit. In any circuit there
conductance) that opposes the flow of curr
a battery (a source of direct current) hooke
an ammeter in the circuit to measure the a
have a circuit as illustrated in Figure 1-17.

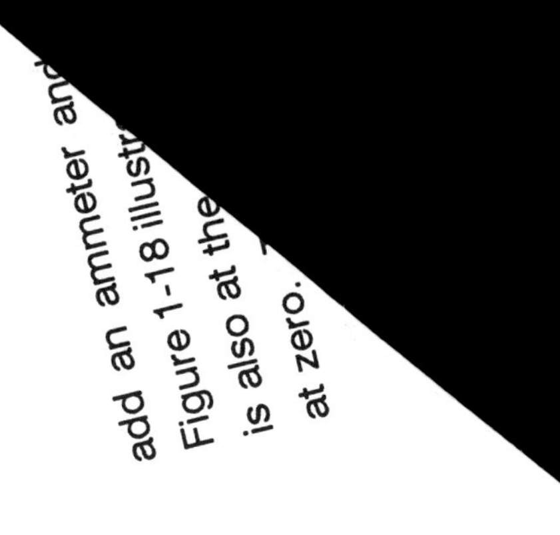

As soon as the *direct current* through the coil
there is no *inductive reactance* from the coil an
the flow of current is the resistance of the wire.

We can compute the amount of resistance in the circuit in Figure 1-17 from Ohm's Law which states that the resistance (R) in a circuit is equal to the voltage (V) divided by the current (I). If we know the voltage of the battery is 12 volts and the meter tells us how much current is flowing, we can compute the resistance in the circuit.

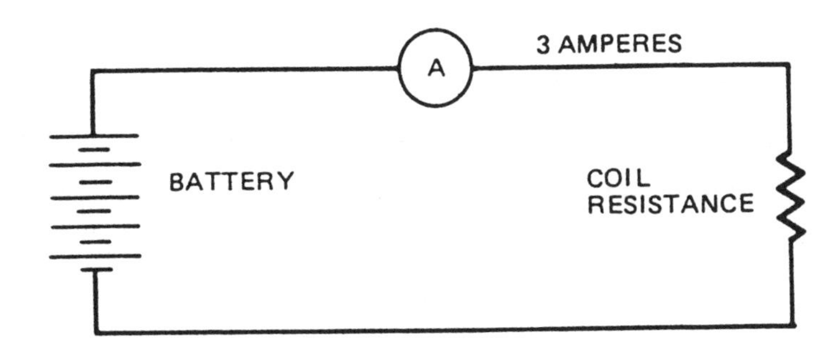

Figure 1-17
DC-Powered Resistive Current

As shown in Figure 1-17, 12 volts divided by 3 amperes equals 4 ohms. The total resistance of the circuit is 4 ohms.

Now that we have discussed how voltage, current, and resistance are interrelated, let's hook up the resistance to a source of *alternating current*, voltmeter as we did before, and plot the results. ...ates that when the voltage is at the maximum, the current ... maximum; and when the voltage is zero, the current is also ... There is no leading voltage or lagging current.

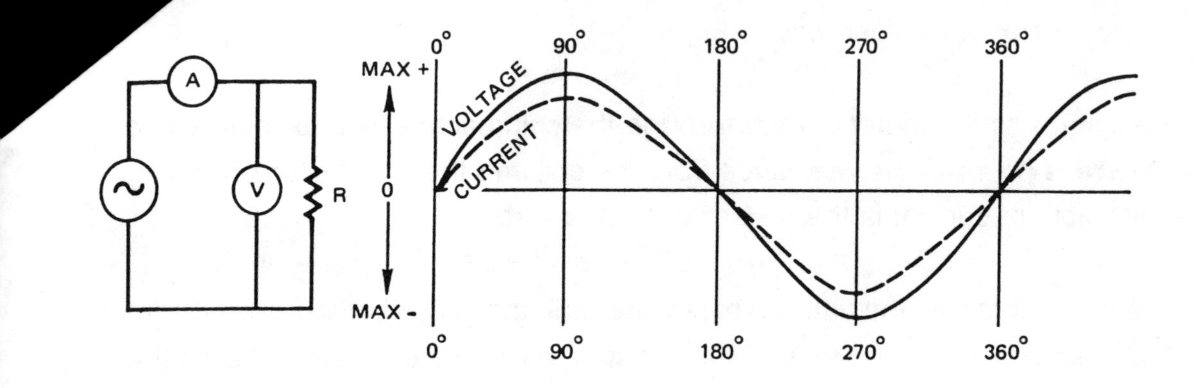

Figure 1-18
AC-Powered Resistive Circuit

In other words, the applied voltage and the resultant current are exactly *in phase* with each other through a resistance. This type of circuit can generate considerable heat when passed through filaments of certain materials. Toasters, light bulbs, and heater coils operate in this way. These coils do not rate as efficient eddy current coils.

In Figure 1-19, we can see that the resistance (R) we found in the circuit is still there. It resists the flow of alternating current just as it did the flow of direct current. The factor that has been added is the inductive reactance of the coil. The inductive reactance (indicated by the letters X_L) causes the current to lag behind the voltage by 90°, i.e., out of phase by 90°.

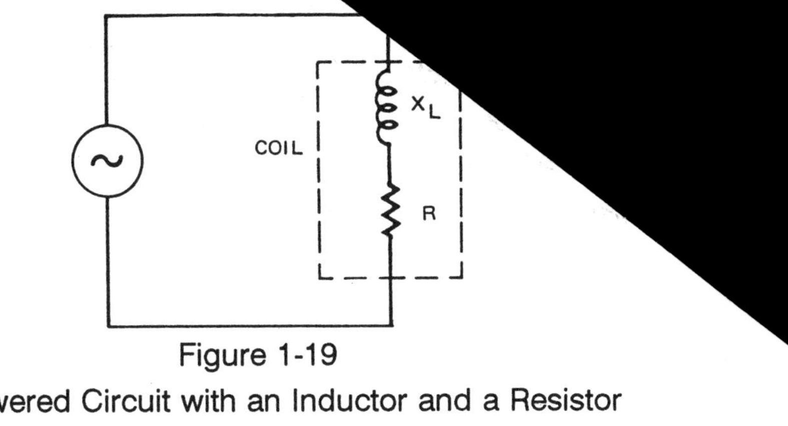

Figure 1-19

AC-Powered Circuit with an Inductor and a Resistor

In an AC circuit, the combination of resistance and inductive reactance is called *impedance* (designated by the letter Z). When we speak of the impedance in an alternating current circuit we mean the *total opposition to current flow* through the circuit and we are including both resistance and inductive reactance.

The total impedance (Z) of the circuit is a combined result of both the resistance (R) and the inductive reactance (X_L). However, the two cannot be added directly because their effect on the voltage is out of phase. The maximum current due to resistance does not occur at the same instant that the maximum current due to inductive reactance occurs.

Here in Figure 1-20 we show a plot of both the current due to resistance and the current due to inductive reactance.

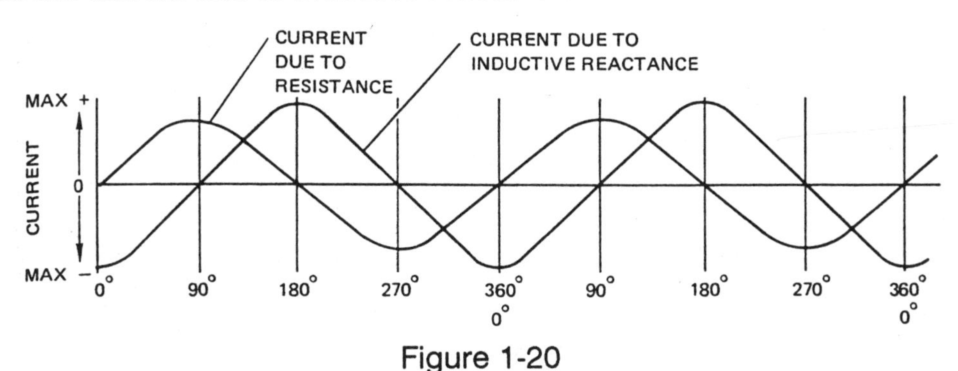

Figure 1-20

Current Affected by Resistance and Inductive Reactance

npedance is the algebraic sum of the
ductive components, we can plot the
ding them together. The results are as
sary to add instantaneous values of both
e resistive component is at its maximum,
zero. The impedance curve must pass
e value since at that point the inductive value

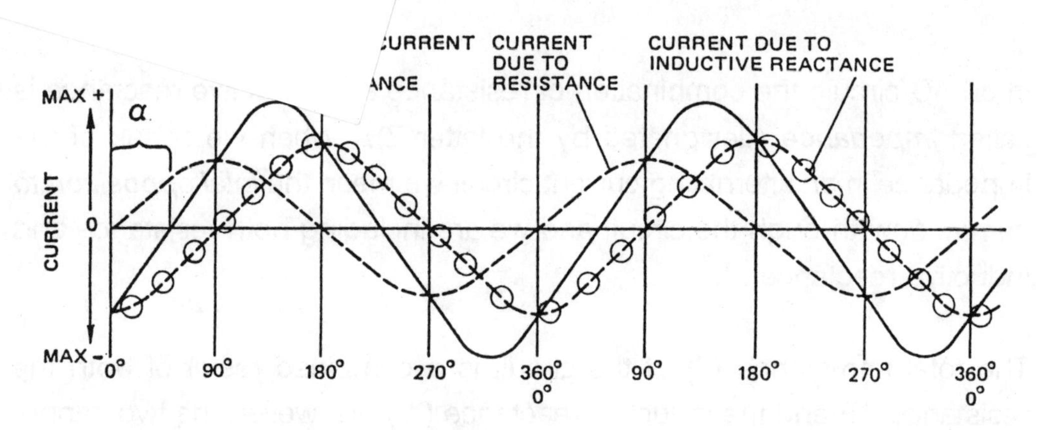

Figure 1-21

Current Flow Based on Total Circuit Opposition

Again, as illustrated in Figure 1-21, at 180° the resistive factor is at zero while the inductive factor is at its maximum. Therefore, the impedance curve must pass through the inductive factor. Note that at any point along the impedance value graph, its value is equal to the sum of the values of current due to both resistance and inductive reactance at that same instant in time.

Note also in Figure 1-21 that I_Z now lags behind the voltage by some angle (α) other than 0° or 90°(in this case, approximately 52°). The voltage and current are said to be out of phase by approximately 52°. Obviously, determining the value of I_Z in this fashion is a long, slow process. Since we are interested only in the maximum current values and the phase lag angle, there is a much easier way to combine these values to obtain the required results.

The simplest way to combine the resistance and inductive reactance values to obtain the impedance value is through *vector addition*. A vector is a line whose length represents a value and its direction represents its phase relationship. We therefore can show resistance and inductive reactance by two vectors 90° apart in direction as illustrated in Figure 1-22, (A). To add these two vectors together, we construct a rectangle then draw a diagonal from corner to corner as shown in Figure 1-22, (B). This diagonal represents *impedance*.

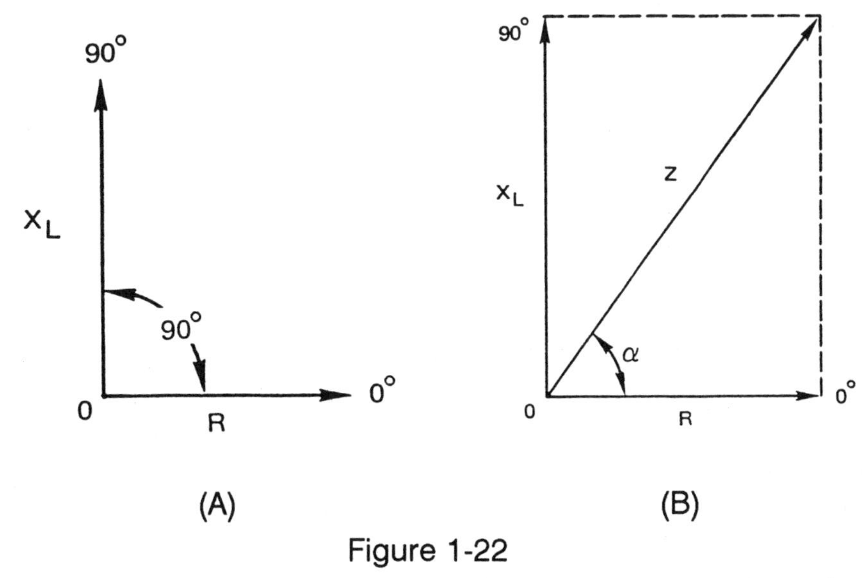

Figure 1-22
Vector Diagram of Resistance, Inductive Reactance and Impedance

The next question is, Is there a way to figure the length of the diagonal mathematically? The answer is yes. A Greek philosopher named Pythagoras figured it out in the sixth century B. C. He developed what is now known as the Pythagorean Theorem which states that in a right triangle the square of the hypotenuse is equal to the sum of the squares of the other two sides. In our situation, this means that $Z^2 = R^2 + X_L^2$. Let's see if that works in our example shown in Figure 1-23.

While you may find this concept interesting, you will probably never need to know this as a eddy current testing technician.

$$Z^2 = 6^2 + 8^2$$

$$Z^2 = 36 + 64 = 100$$

$$Z = \sqrt{100} = 10 \text{ ohms.}$$

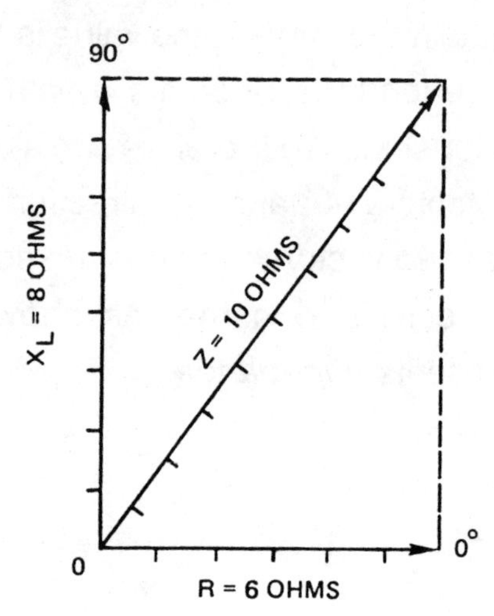

Figure 1-23
Pythagorean Theorem

The concept that you must fully understand is the concept of what is meant by "phase angle." In the illustrations on the last few pages, the phase angle between the resistance vector and the inductive reactance vector has been 90°. In Figure 1-24, the Greek letter alpha (α) is used to denote the phase angle of the *impedance* vector.

The area that we want to explore next is the effect that changes in the circuit parameters have on the phase angle. First, let's look at resistive changes. With all other factors held constant, changes in resistance will affect the impedance of the circuit. Figure 1-24 shows that as the resistance value increases, the impedance value increases and the phase angle *decreases*.

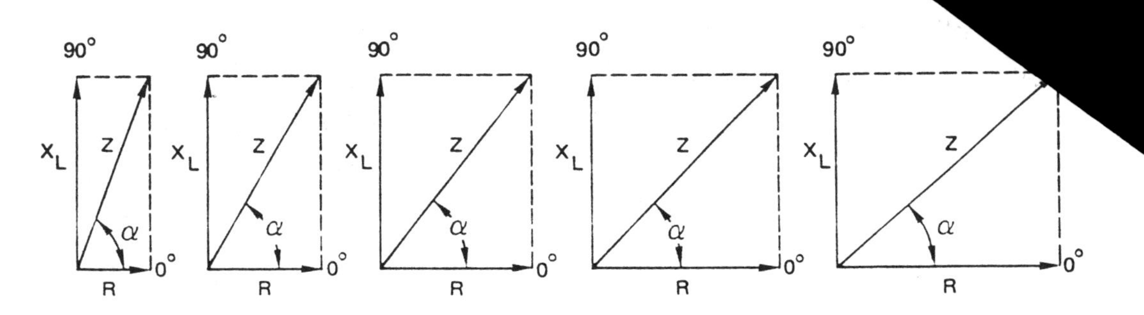

Figure 1-24
Variations in Resistance Affecting Impedance

Now, here's an illustration in Figure 1-25 that shows what occurs when the inductive reactance changes while all other factors are held constant. As the inductive reactance increases, the phase angle of the impedance increases. So it approaches (moves toward) 90°.

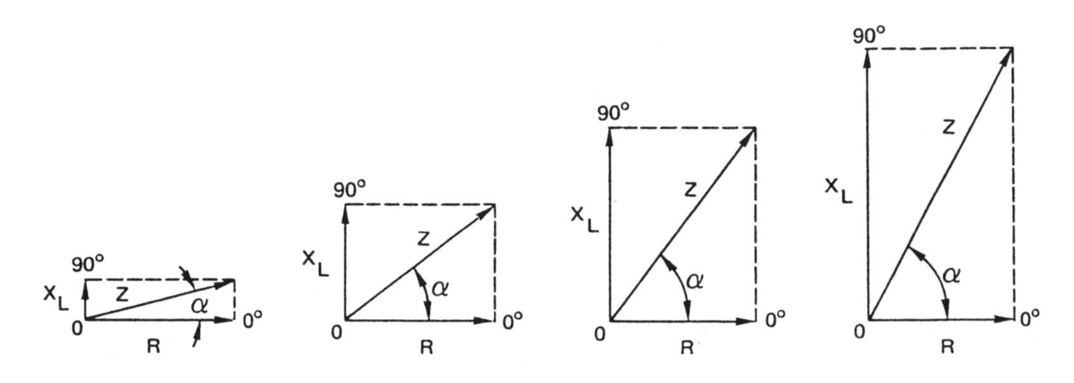

Figure 1-25
Variations in Inductive Reactance Affecting Impedance

In a given alternating current circuit, the resistance value stays fairly constant. The value can change if the temperature of the wiring increases or decreases but the change is slight. **Changes in voltage or frequency do not affect the resistance that is in the circuit.**

e, however, *is* affected by frequency changes. Any given
ular value of inductance (L) measured in Henries. (The
coil is one Henry when a current variation of one ampere
luces one volt.) The inductance value is determined by the
and number of turns in the coil and is, therefore, constant for
any oil.

The inductive reactance (X_L) in ohms is determined by the equation $\mathbf{X_L = 2\pi fL}$. For any given coil the terms 2, π, and L are constant; thus, the inductive reactance for any given coil depends entirely on the frequency of the current through the coil. Increasing the frequency causes the inductive reactance to increase proportionally.

Doubling the frequency will cause the inductive reactance to double. Let's see what this looks like on a graph so we can see what happens to the impedance of the circuit. In Figure 1-26 we are showing what happens as the result of changing frequency. As you can see, if the frequency were doubled the phase angle of the impedance would increase.

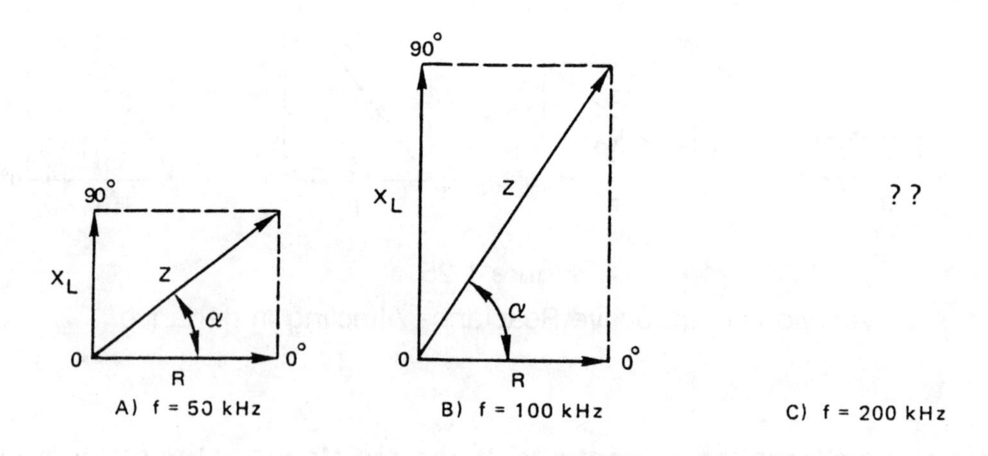

Figure 1-26
Variations in Frequency Affecting Impedance

We have now covered the theory of electromagnetic induction to a depth sufficient for you to be able to understand its applications to eddy current testing. **To summarize**:

- Electromagnetic induction is the process of generating an electrical current in another conductive material by varying the way the primary flux density interacts with the secondary material.

- The primary circuit impedance, or the *total opposition* to current flow in this coil, is a combination of two variables:

  - the *DC resistance* of the wire and the *inductive reactance* of the coil.

- Since the DC resistance will not vary significantly, it is the coil's inductive reactance that will most affect its efficiency and operation.

- The "variables" in our primary circuit affecting the inductive reactance are:

  - the number of turns of wire in the coil.
  - the physical size of the coil.
  - the amount of current in the coil.
  - the frequency of the current supplied to the coil.

- How strong our secondary current will become will be determined by the electrical conductivity of the secondary material that we are attempting to generate this new current in.

CHAPTER 2

PRINCIPLES OF EDDY CURRENT TESTING

In Chapter 1 we discussed the results that Faraday achieved in 1831 with electromagnetic induction between two coils. Additional scientific advances were required before electromagnetic induction could be applied as an inspection process. In 1879, Hughes detected differences in electrical conductivity, magnetic permeability, and temperature in metals using an eddy current method. The eddy current method developed slowly, partly because such a method was not needed. By 1935, however, eddy current testing was being done.

It is unfortunate that we cannot use our senses to witness what eddy currents look like or how they work. We have to rely on visual representations and models. Hopefully, these models can help us to understand what is going on during eddy current testing.

In Figure 2-1, we have the following situation:

- Conductor A represents a small part of a coil winding.
- Conductor B represents our test sample.
- The linear and circular arrows represent the invisible electrical and magnetic fields in the two conductors.

- It is important to note both the direction and the size of these arrows or "vectors" and how they relate to each other.

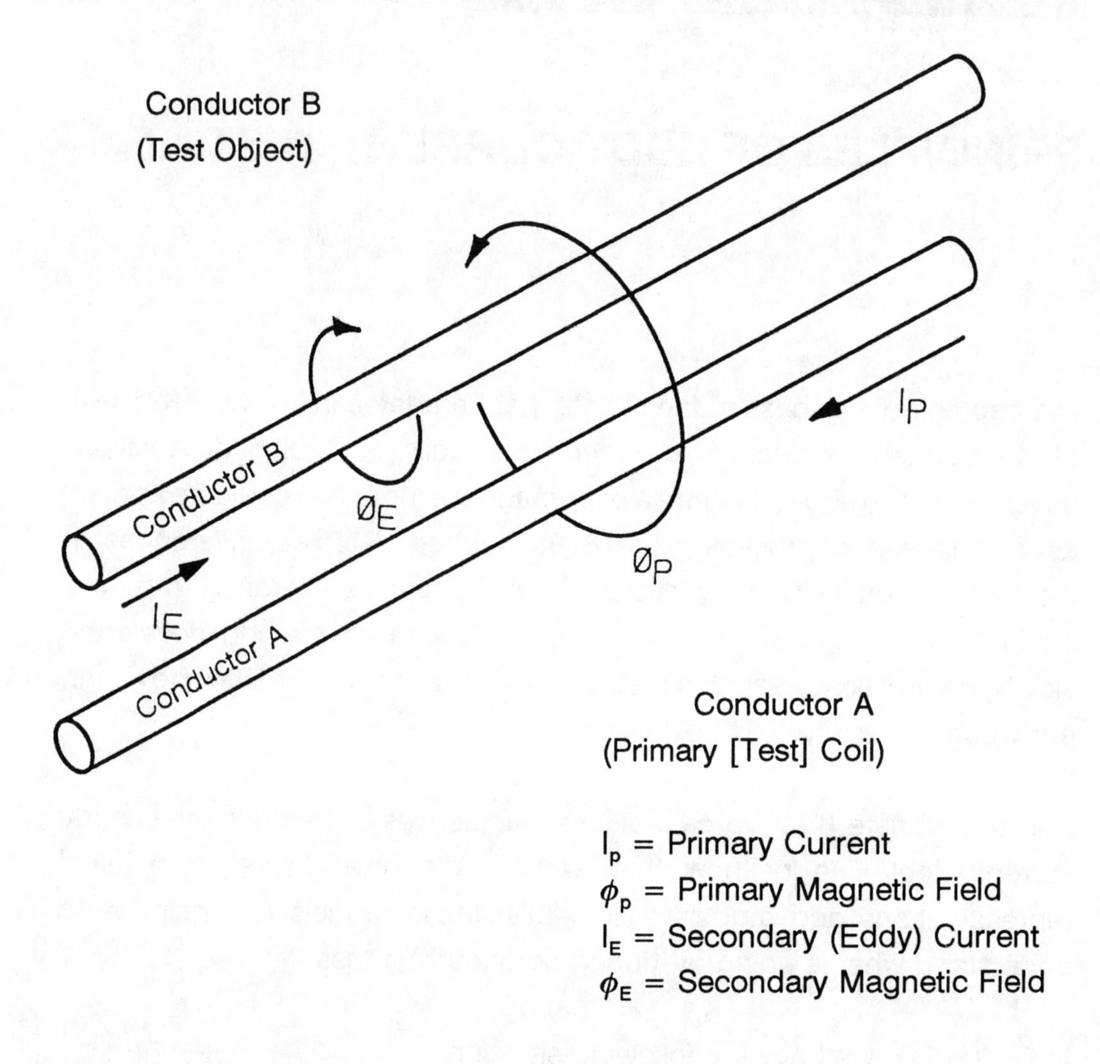

Figure 2-1
A Visual Representation of Electromagnetic Interactions

Note in Figure 2-1 that the Primary Current (I_P) is much larger than the Secondary (Eddy) Current (I_E) in the test specimen and that it is moving in the **opposite direction.**

Note also that the Primary Magnetic Field (ϕ_P) is much larger than the Secondary Magnetic Field (ϕ_E) and these two forces are also acting in **opposite directions.**

The last important feature to remember about Figure 2-1 is that in eddy current testing we are using an AC power source probably operating in the thousands or even millions of cycles per second range. So this figure is like a high speed photograph. We are looking at a **single moment in time.** In the next fraction of a second, the entire picture will change as shown in Figure 2-2.

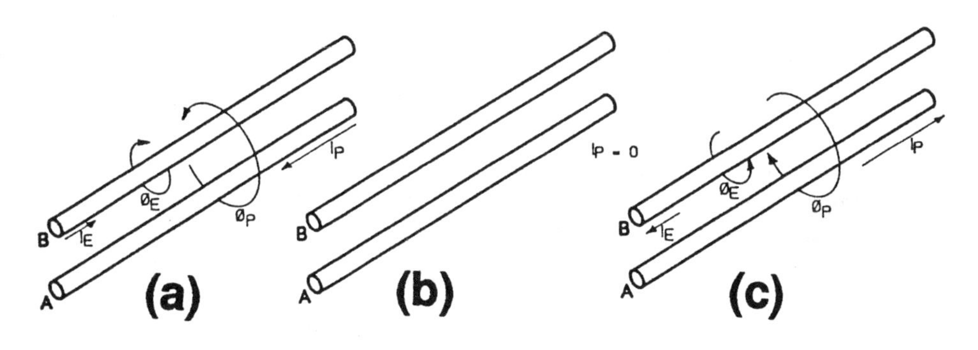

Figure 2-2
Changes in Magnetic Forces as Current Changes

Fig. 2-2 (a) Primary current at maximum creating responses ϕ_P, I_E, and ϕ_E as shown.

Fig. 2-2 (b) Primary current passes its zero point. All of the associated responses are "off."

Fig. 2-2 (c) Primary current passes its zero point and begins to build in the OPPOSITE direction to a new maximum. Vectors ϕ_P, I_E, and ϕ_E will reach their maximums but in the opposite direction as those shown in Figure 2-3.

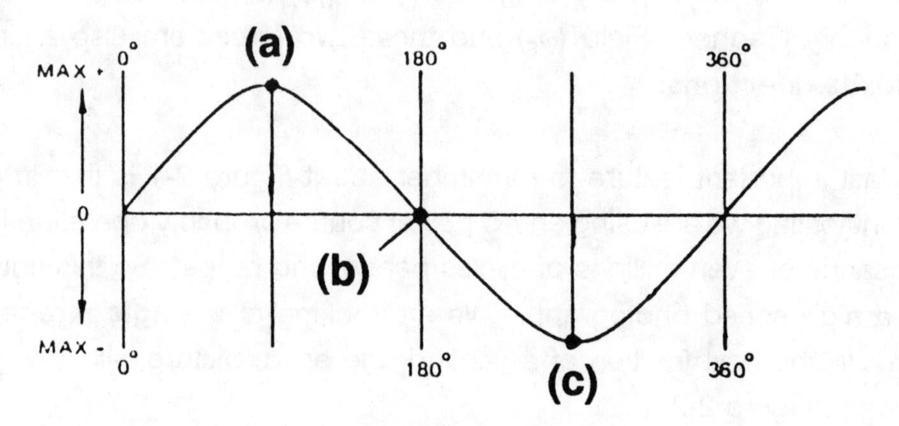

Figure 2-3
Changes in Magnetic Forces as Current Changes

Faraday's electromagnetic induction principles are applied in eddy current testing as follows:

- A coil is energized by an alternating primary current, creating an alternating magnetic field around it.
- This inspection coil is placed on or near a test specimen.
- If the test specimen is a conductor of electricity, *small circulating electrical currents* (eddy currents) are induced in the specimen.
- These eddy currents produce a secondary magnetic field that opposes the primary magnetic field causing the eddy currents.
- The opposing secondary magnetic field causes a change in the original magnetizing field.
- Changes in the primary magnetic field cause a change in the impedance of the coil.

- The change in impedance can be detected.

The important concept here is that the article under inspection *must* be capable of conducting electrical current.

Let's now look at the relationship between the magnetic field of an inspection coil and the material under test. We know that the material must be a conductor of electricity. In Chapter 1, we learned that when a magnetic field cuts through an electrical conductor, current is induced in the conductor. Of course, the conductor was a wire or a coil. Let's see what happens if the conductor is a thick sheet of *electrically-conductive* material.

As shown in Figure 2-4, an electrical current is induced in the material. This current flows in a *circular path (eddy)* and, like the current in the coil, produces a magnetic field. The magnetic field created around the eddy current is *at all times* in the opposite direction to the magnetic field of the coil.

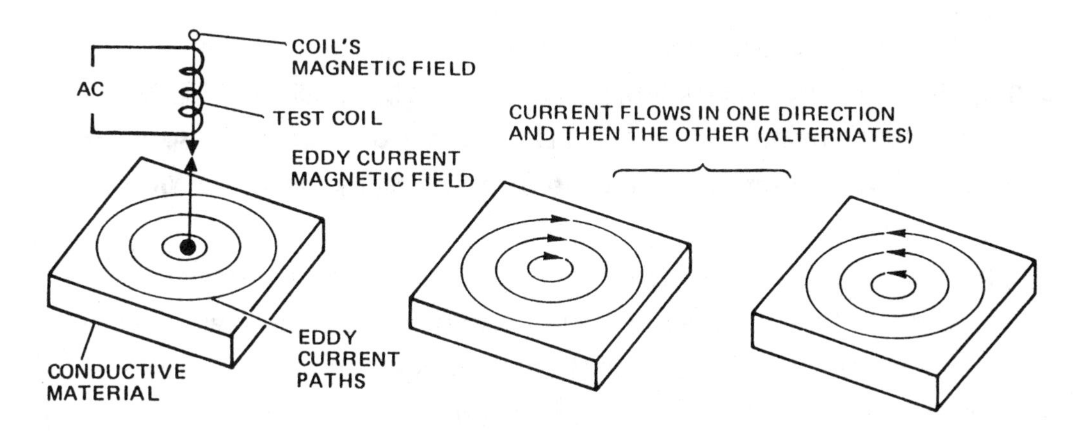

Figure 2-4
Induced Eddy Current Flow in a Conductive Plate

The action that we can't show clearly in this illustration is the fact that since the current flow in the inspection coil is alternating and changing its direction, the magnetic field is also continually changing its direction, and the eddy currents and their magnetic field are also continually changing their direction, as shown in Figure 2-5.

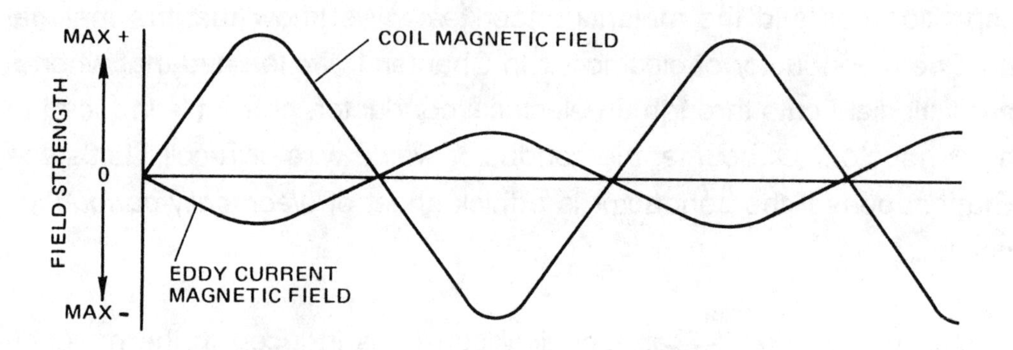

Figure 2-5
Opposition of Primary and Secondary Magnetic Forces

In spite of the fact that the currents and magnetic fields involved are constantly changing (alternating direction) the phase relationship between each always remains the same. **Thus the eddy current secondary magnetic field always opposes the primary coil's magnetic field.**

The next factor to consider is the strength of the eddy current magnetic field. All other things constant, the strength of the secondary magnetic field depends directly on the strength of the eddy currents. Stronger eddy currents cause a stronger secondary magnetic field which opposes the coil's magnetic field.

We said that one factor that affects the amount of current induced in a secondary coil is the conductivity of the material of which the coil is made. We said that more electric current will be induced in a coil with high conductivity than in a coil with low conductivity. The same holds true for the induction of eddy currents in articles to be inspected by the eddy current method. With all other factors held constant, stronger eddy currents will be induced in articles that have higher conductivities than will be induced in articles having lower conductivities.

Here in Figure 2-6 we show *a coil* being used to inspect two articles that are just alike except that one is made of copper and the other is made of stainless steel.

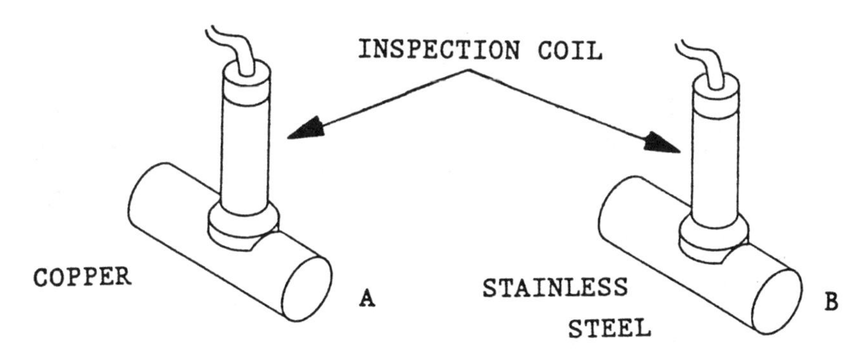

Figure 2-6
Measuring Material Differences in Bar Stock

Since copper is a better conductor of electricity than stainless steel, stronger eddy currents will be induced in article A than in article B. These stronger eddy currents will result in a stronger secondary magnetic field to oppose the magnetic field of the coil. Remember then that the *conductivity* of the material is one factor that affects the induction of eddy currents in the material.

Now let's look at another factor which affects the eddy currents and the secondary magnetic field. Consider the magnetic field that surrounds the inspection coil as shown in Figure 2-7. The field extends from each end of the coil. The field is at its strongest next to the coil and progressively gets weaker away from the coil.

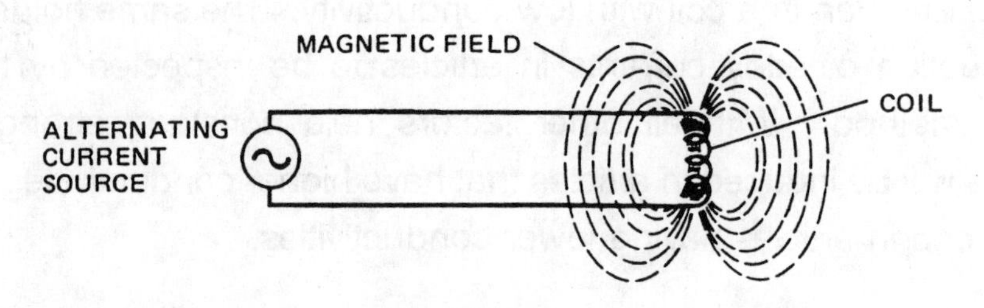

Figure 2-7
Magnetic Field About an Energized Coil

Consider now a comparison between the strength of the coil's magnetic field and the strength of the induced secondary magnetic field. The stronger the coil's magnetic field, the stronger the secondary magnetic field opposing it. Since the coil's field is strongest near the coil and gets progressively weaker away from the coil, we can hold the coil a short distance away from conductive material and still induce eddy currents in the material as shown in Figure 2-8.

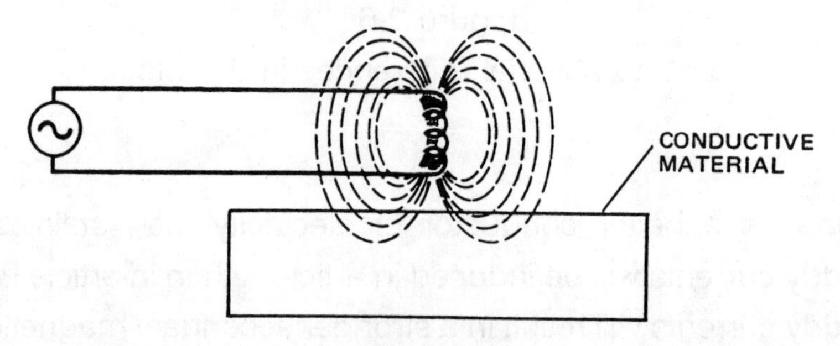

Figure 2-8
Coil's Magnetic Field Affecting a Test Sample

The eddy currents induced in the conductive material will be much weaker than the eddy currents induced if the inspection coil were placed directly on the part, but so long as the conductive material intercepts any of the primary coil's magnetic field there will be a corresponding amount of eddy currents induced in the material.

Now let's see how we can use this fact in eddy current testing. As shown in Figure 2-9, we can cover the conductive material with a coating of *nonconductive* material and place the test coil directly on the nonconductive coating. The coil's magnetic field is unaffected by the presence of the nonconductive coating on the material. The same field exists but only the weaker portion is effective in inducing eddy currents in the conductive material.

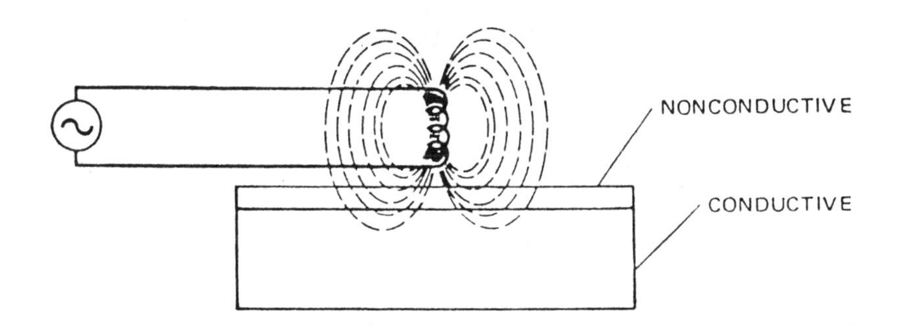

Figure 2-9
Coil's Magnetic Field Affecting a Nonconductive Coating

By making the nonconductive coatings thinner, we have moved the inspection coil closer to the conductive material. Since the coil is now closer to the conductive material, the eddy currents will be stronger.

Air is a nonconductor of electricity. If the inspection coil is lifted even slightly from the surface of the article being inspected, the magnetic fields in the article are affected. The magnetic fields change. These changes can be mistaken for changes caused by other factors. Being able to differentiate changes caused by this "lift-off" from changes caused by other factors is an important part of the science of eddy current testing. "Lift-off" is the name given to this factor.

Lift-off is not a factor that results from something in the material. It results from the inspection procedure. Lift-off has been put in the category of being a *dimensional* factor. There is another dimensional factor that does result directly from the material. The illustration in Figure 2-10 shows this other dimensional factor.

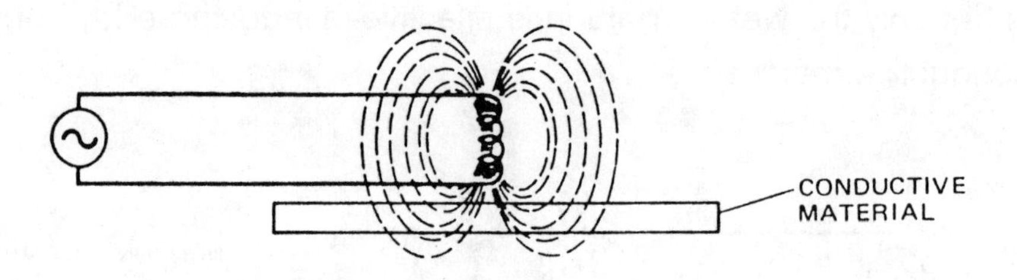

Figure 2-10
Position of Flux Lines as Material Becomes Thinner

If we put the inspection coil on progressively thinner and thinner pieces of conductive material, we reach a point where the material is intercepting only part of the coil's magnetic field. In Figure 2-10 we see that the primary magnetic field extends all the way through and beyond the back surface of the test specimen. Thus the induced eddy currents must be reduced. Since the eddy current strength is reduced, the secondary magnetic field is also reduced and the magnetic field through the coil is changed.

In Figure 2-11 we show two different thicknesses of copper being tested with an inspection coil.

We cannot know whether different results will be obtained until we know the actual thicknesses. *If at least one* of the two specimens were thin enough to not contain the entire magnetic field of the coil, then we would get different results from the two tests.

Remember, if the materials are thick, minor differences in thickness have little or no effect on eddy current distribution.

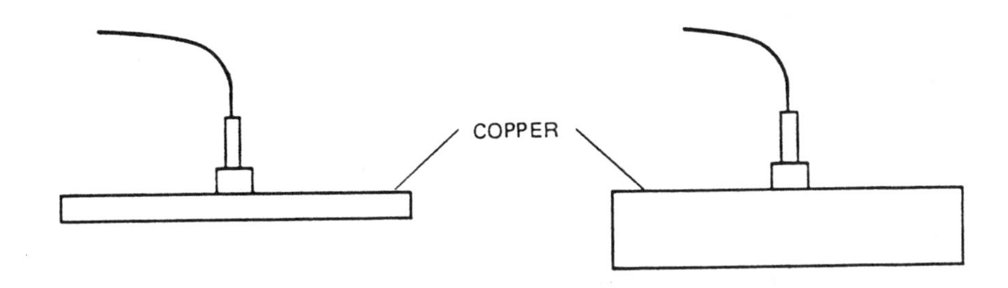

Figure 2-11
Variations in Material Thickness

So now we have two factors, lift-off and material thickness, that are DIMENSIONAL factors to be considered in eddy current testing.

We know that CONDUCTIVITY factors and DIMENSIONAL factors affect the results of eddy current tests. The third, and last, major factor is the magnetic PERMEABILITY of the material. *Some* conductive materials are also permeable, which means that they can be magnetized.

When the material under test is ferromagnetic (has permeability), the coil's magnetic field is affected in an inconsistent way. For example, if the coil's magnetic field is relatively weak, the slightest variation in the induced eddy current may cause relatively large changes in the coil's magnetic field. On the other hand, if the coil's magnetic field is relatively strong to begin with, slight variations in the eddy current field may cause only slight variations in the coil's magnetic field. As you can probably understand, this type of inconsistency in the test results would be difficult to live with.

Let's take a piece of unmagnetized soft iron and magnetize it by placing it in a coil to which we can apply direct current. Let's start with zero current through the coil, then gradually increase the current. If we measure the strength of the magnetic field induced in the iron during the time we increase the current, we will find that the magnetic field strength increases sharply at first, then at the higher levels of current the strength of the magnetic field ceases to increase regardless of any increase in magnetizing current. If we plot current values against magnetic field strength we get a curve as shown in Figure 2-12.

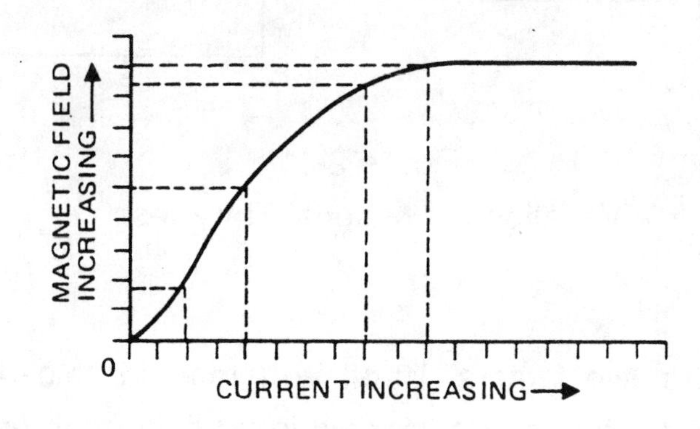

Figure 2-12

Magnetizing Force Plotted Against Magnetic Field

Also note in Figure 2-12 that for lower values of current (the left-hand part of the curve) two units of current change caused about three and a half units of change in field strength while at the higher values of current two units of current change caused less than one unit of change in field strength. It is this variation in test results that we cannot live with. At the right end of the curve the soft iron has become magnetically saturated as shown in Figure 2-13. Changes in the magnetizing current no longer have any effect on the strength of the induced magnetic field.

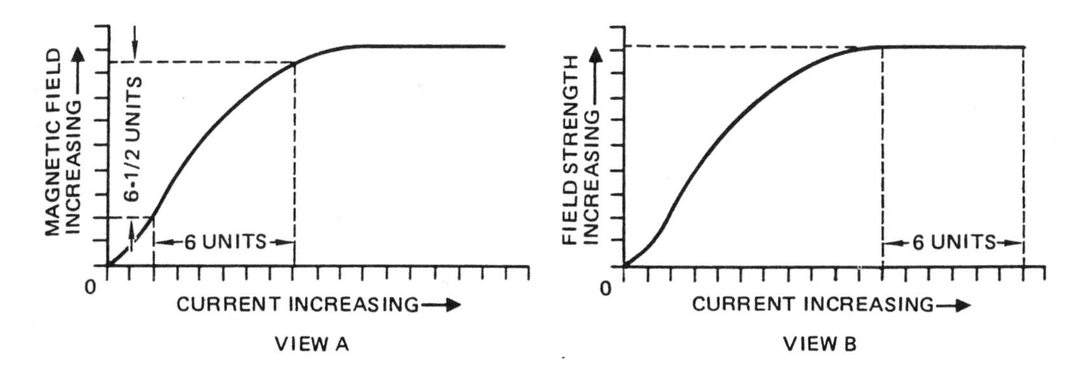

Figure 2-13

Magnetizing Force Plotted Against Magnetic Field

This gives us a clue to a method that can be used to overcome the problems caused by the PERMEABILITY factor. Suppose that we use a separate coil (not an inspection coil) powered by direct current to magnetize the test specimen to its saturation point. When the inspection coil is then applied, the eddy currents induced are affected only by the CONDUCTIVITY and DIMENSIONAL factors of the test specimen. The *inconsistencies* caused by the permeability factor have been eliminated. Note that we have not eliminated the permeability, we have eliminated the inconsistencies caused by the fact that the material is ferromagnetic.

It is very difficult to visualize the interaction of the magnetic fields involved when testing ferromagnetic materials. But, if you remember that eddy currents are induced only by changing magnetic fields, the situation is easier to visualize.

For both surface scanning and thin-walled tubing inspection, it is possible to mount magnets near the inspection coils in the same assembly, as shown in Figure 2-14. This allows the DC saturation field to travel at the same rate of speed as the eddy current coils. As long as the DC magnetic field has the same relative field strength in the eddy current inspection zone, then there will be no interference between the two fields. This technique can compensate for minor variations in the sample permeability and improve the inspection output signal consistency.

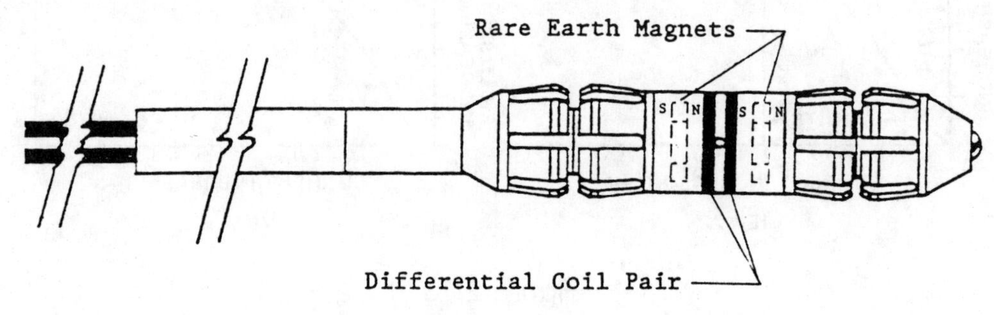

Figure 2-14
Magnetic Bias Bobbin Probe

As illustrated in Figure 2-15, it is necessary to use large, stationary DC saturation coils (or yokes) for the inspection of materials with a high permeability rating (ferritic tubing, rod, wire, etc.). This is normally done in the production plant environment. Once these materials are put into assemblies, it becomes much more difficult to achieve the levels of saturation required to overcome their inherent permeability.

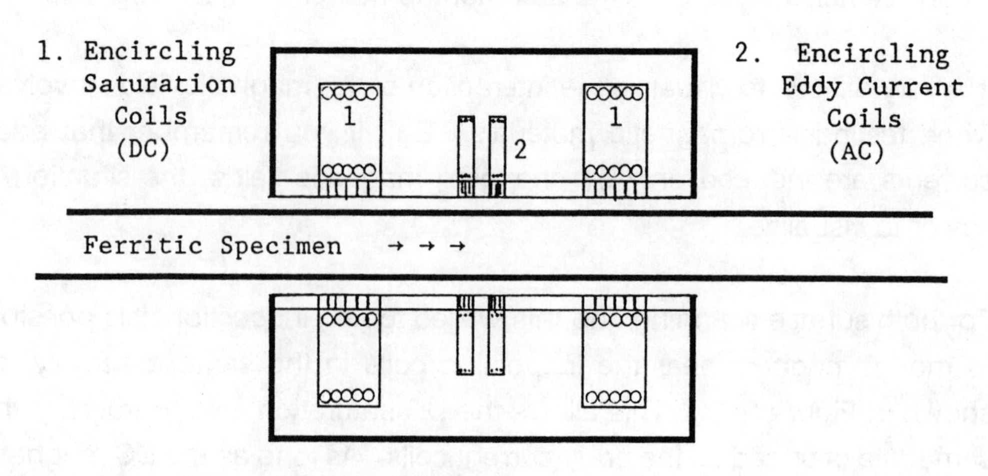

Figure 2-15
Typical Ferritic Tubing Mill EC Inspection Assembly

Now let's take a look at conductivity. Conductivity is a measurement of the ability of a material to carry electrical current. The conductivity of a given material is defined as the number of amperes of current that will flow through a given size (cross-sectional area) of the material when a given voltage is applied to it. This definition is too cumbersome to use when applied to eddy current testing. Instead, we use the International Annealed Copper Standard (IACS) of conductivity.

In the IACS the conductivity of unalloyed (pure) annealed copper was arbitrarily selected as the standard and the conductivities of all other materials are expressed as a percentage of this standard. In other words, annealed copper is assigned a rating of 100 percent and a material that conducts electrical current only half as well is rated at 50 percent IACS. The IACS rating of conductivity for several common materials is given in a table in Figure 2-16.

METAL OR ALLOY	percent IACS	METAL OR ALLOY	percent IACS
SILVER	105	70-30 Cu-Ni	4.6
COPPER, ANNEALED	100	PHOSPHOR BRONZE	11
GOLD	70	MONEL	3.6
ALUMINUM	61	ZIRCONIUM	3.4
ALUMINUM ALLOYS:		ZIRCALOY-2	2.4
6061-T6	42	304 STAINLESS STEEL	2.5
7075-T6	32	INCONEL 600	1.7
2024-T4	30	HASTELLOY X	1.5
MAGNESIUM	37	WASPALOY	1.4

Figure 2-16
International Annealed Copper Standard (IACS) Ratings

This table is only a partial list of materials, but it illustrates that materials do have different abilities to conduct electrical current. They range from good conductors, like copper and silver, to poor conductors, like nickel or steel, to nonconductors, like wood or glass.

Now, if we establish a given magnetic field in an inspection coil and place the coil first on a piece of copper and then on a piece of aluminum, the coil's magnetic field will be affected differently by each metal because of the difference in conductivity. The stronger eddy current induced in the copper will produce a stronger secondary field in opposition to the field of the coil. Notice that we used a given magnetic field to test those materials. If we had used a different magnetic field in each of those two tests we would not have been comparing the two metals under exactly the same conditions. The results would have varied accordingly.

The point we are making is this: differences in the conductivity of different materials is detectable by eddy current testing due to the effect that the conductivity of the material has on the magnetic field of the coil.

A statement can be made about materials in terms of conductance or resistance. For example, a good conductor must be a poor resistor. Unfortunately, we cannot directly compare values of conductivity and resistivity because they have different origins and units.

- The symbol for conductivity is σ (sigma) and the units in this case are expressed in percent IACS. Keep in mind that this is based on an arbitrary standard chosen for convenience and addresses how well a material permits current flow.

- The symbol for resistivity is ρ (rho) and the units are expressed in micro-ohm-centimeters ($\mu\Omega$cm). Keep in mind that this is based on an *absolute* scale and addresses how a material resists current flow.

We have stated that a good conductor must, by default, also be a poor resistor. Let's look at a couple of examples taken from the table in Figure 2-16 before we continue.

The material shown in Figure 2-17 demonstrates two important facts:

- **As resistivity increases, conductivity decreases.**
- **As conductivity increases, resistivity decreases.**

MATERIAL	RESISTIVITY (ρ) rho in micro-ohm-cm ($\mu\Omega$cm)	CONDUCTIVITY (σ) sigma in % IACS
Copper	1.72	100.00
	↓	↑
Inconel 600	100.00	1.72

Figure 2-17
Comparison of Resistivity and Conductivity Values

In Chapter 6 we will learn to look at a range of material responses and begin to make judgements about their general relationship to one another. If we were to plot the two materials shown in Figure 2-17 on what is commonly called a Conductivity Curve, we would see that they fall at opposite ends of the material graph.

Look at the relationship of the resistivity and conductivity values of the materials shown in Figure 2-17. You can easily see that as one value increases, the other value decreases. These numbers are values of the same material property but address it on a different scale.

Is a glass of water half full or half empty? Is a given point in a material more resistive or less conductive than the material around it? The reference point is still the same regardless of how you ask the question. The question we are asking is how well will eddy currents flow at some point of interest in the test material? The controlling factors are:

- Conductivity (or Resistivity)
- Permeability
- Geometry (Dimensional Parameters)

Since the relative conductivity of metals and alloys varies over a wide range, the need for a conductivity benchmark was of prime importance. The International Electrochemical Commission established a convenient method of comparing one material to another in 1913. The Commission established that a specific grade of purity copper of a given size with a measured DC resistance of **0.017241 ohms** would be arbitrarily considered 100 percent conductive.

Figure 2-18 is an expanded version of Figure 2-17. Notice what happens to the resistivity and conductivity scales.

With the understanding that conductivity and resistivity work in opposition to each other, it is possible to make direct conversions between these scales by using simple division. The original copper bar used to establish the standard had a DC resistance of **0.017241 ohms.** We can disregard most of problems created by trying to keep track of decimal places and the systems of units by using a conversion factor of **172.41.**

$$\text{Conductivity (percent IACS)} = \frac{172.41}{\text{Resistivity } (\mu\Omega\text{cm})}$$

OR

$$\text{Resistivity } (\mu\Omega\text{cm}) = \frac{172.41}{\text{Conductivity (percent IACS)}}$$

MATERIALS	RESISTIVITY MICRO-OHM-CM (μΩcm)	CONDUCTIVITY percent IACS
Ti-6Al-4V	172.00	1.00
Hastelloy-C	132.62	1.30
Hastelloy-X	115.00	1.50
Inconel 600	100.00	1.72
Stainless Steel 316	74.00	2.33
Stainless Steel 304	72.00	2.39
Zircalloy-2	72.00	2.40
Titanium-2	48.56	3.55
Monel	47.89	3.60
Zirconium	40.00	4.30
Copper Nickel 70-30	37.48	4.60
Uranium	≃30.00	5.70
Lead	20.65	8.35
Copper Nickel 90-10	18.95	9.10
Phosphor Bronze	16.00	11.00
Aluminum Bronze	12.32	14.00
Admiralty Brass	6.90	25.00
70-30 Brass	6.20	28.00
Tungsten	5.65	30.51
Aluminum 7075-T6	5.39	32.00
Aluminum 2024-T4	5.20	30.00
Magnesium (99percent)	4.45	38.60
Sodium	4.20 @ 0°C	41.05
Aluminum 6061-T6	4.10	42.00
Aluminum (99percent)	2.65	64.94
Gold	2.46	70.00
Copper	1.72	100.00
Silver	1.64	105.00

Figure 2-18
Electrical Resistivity and Conductivity
of Common Metals and Alloys

If we know the conductivity of a material from an IACS chart, then we can convert that into a number which tells us the material's resistivity. The resistivity value happens to be expressed in something called micro-ohm-centimeters (μΩcm) but we do not need to know that. It is the actual values involved that we may need.

So why on earth do we want to throw in another "variable" at this point? What good is knowing the resistivity of material? In this chapter, we are attempting to show you some simplified formulas that may allow you to do a better eddy current examination. These formulas are going to use combinations of the variables that we have been discussing. The easiest factor for the eddy current technician to control is the operating frequency of the examination. Frequency is very important in determining where the eddy current field exists in a test sample.

Based on further discussions in this chapter, it will be possible to mathematically predict the location and sensitivity of our eddy current field during a test. With this information at our disposal, we may be able to more accurately understand the results that are obtained as we inspect a test sample.

Let's talk about depth of penetration of eddy currents. We said earlier in this chapter that the magnetic field of a surface coil may not penetrate entirely through a thick material. We gave no indication as to what the thickness limit might be nor did we mention any of the factors that affect the depth of penetration. We shall do so now.

First, you must realize that eddy currents are not uniformly distributed throughout an article being inspected. They are most dense at the surface immediately beneath the coil and become progressively less dense with increasing distance below the surface. Figure 2-19 shows the relative distribution of eddy currents in any material. The numbers up the left side of the graph give the relative density of the eddy current as a percentage of the density of the current that exists at the surface. The numbers across the bottom give the material depth as multiples of the "standard depth of penetration."

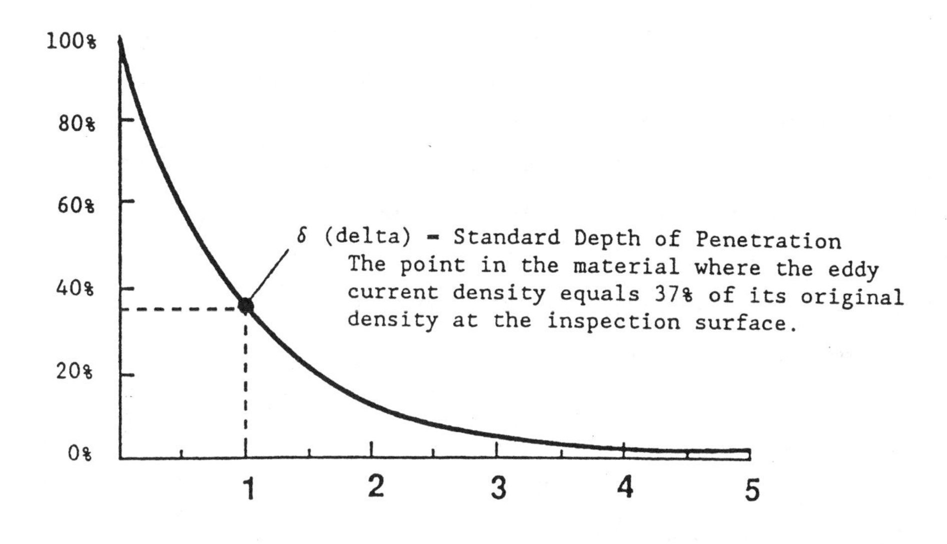

Figure 2-19
Graph Showing Standard Depth of Penetration

The standard depth of penetration is defined as the depth at which the current is approximately 37 percent of the current density that exists at the surface. If the standard depth of penetration exceeds the thickness of the material under test, the restriction of the eddy current paths appears as a change in the conductivity of the material. The coil response then reflects the thickness of the material. It should be remembered, however, that eddy currents do not cease to exist beyond the standard depth. Normally the material must have a thickness of two or three times the standard depth of penetration before thickness ceases to have any effect on the test coil.

Of course it is reasonable to expect that the standard depth of penetration would vary between different materials, which it does. The depth of penetration is affected by the *conductivity* and the *permeability* of the material.

Let's see why conductivity affects the depth of penetration. You recall that the magnetic field produced by the eddy currents opposes the magnetic field of the coil. You also recall that induced eddy currents are stronger in the more conductive materials. Then it is reasonable that these stronger eddy currents will produce a stronger secondary magnetic field with stronger opposition to the coil's magnetic field. Thus the coil's magnetic field is considerably weakened near the surface. This effect continues as the coil's magnetic field enters deeper into the material. At every level it is met with strong opposition.

We want to emphasize that the conductivity of the material and the permeability of the material are the two factors related *to the material* that affect the depth of penetration.

THE HIGHER THE CONDUCTIVITY, THE LESS THE PENETRATION.

THE HIGHER THE PERMEABILITY, THE LESS THE PENETRATION.

There are two other considerations that affect the depth of penetration that are controllable to a degree by the operator. These are the geometry of the exciting coil and the frequency of the power supply.

The effect of the geometry of the exciting coil may be visualized by imagining the comparison of a closely wound coil versus a loosely wound coil. There is bound to be a difference in their penetration qualities due to the difference in their magnetic fields. It is difficult to draw any hard and fast rules that govern the effect of coil size, etc., since there are so many variables. We will address some factors of coil design and their applications in the following chapters.

A third factor that is controllable by the operator (on some eddy current test equipment) is the frequency of the alternating current applied to the coil. When a higher frequency is used, the magnetic field is changing more rapidly. As we pointed out in Chapter 1, a higher rate of change of the magnetic field will cause higher values of current to be induced in the material. Stronger eddy currents cause a stronger secondary magnetic field that opposes the penetration of the coil's magnetic field.

The lower the frequency, the greater the depth of penetration. We have discussed three factors that affect the depth of penetration. Two are related to the material itself and the third is controlled by the test equipment operator.

Figure 2-20 shows the standard depth of penetration that results for several materials at different operating frequencies. The graph also indicates the effect of conductivity and permeability on the depth of penetration. For example, note that the depth of penetration in copper is less than the depth of penetration in aluminum at any given frequency. This is because copper is a better conductor of electricity than aluminum.

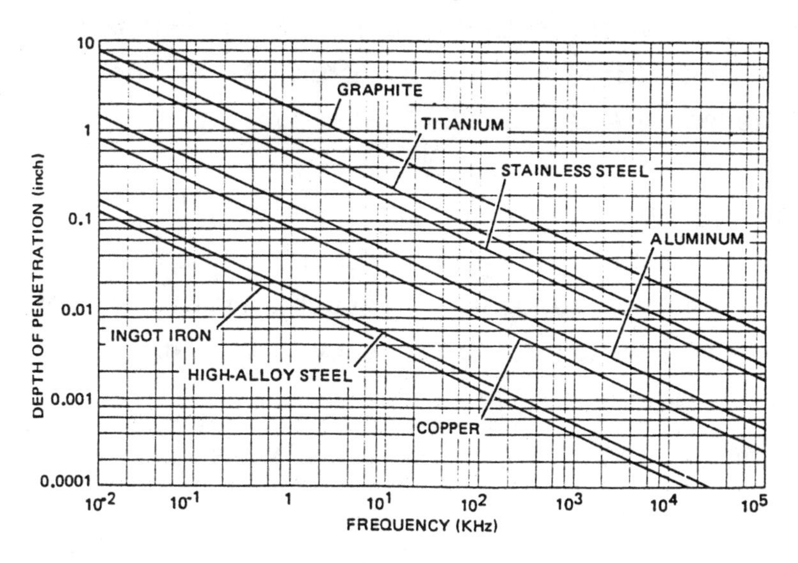

Figure 2-20

Graph of Frequency Versus Depth of Penetration

Let's assume that the conductivity of high-alloy steel and iron is almost the same. If this is true, then the difference in depth of penetration must be due to the difference in permeability. Therefore, the iron is the more permeable, since the depth of penetration in iron is less than the depth of penetration in high-alloy steel at any given frequency. Since the depth of penetration in iron is less, it must be more permeable. Note also the large effect that permeability has.

Graphite, titanium, stainless steel, aluminum and copper are all nonferromagnetic materials so they have no permeability. The iron and high-alloy steel are certainly not better conductors than copper, yet they appear below copper on the graph. Thus their permeability has caused loss of penetration depth beyond even one of the best conductors.

What happens if we are inspecting a material that doesn't show up on a standard depth of penetration chart? How do we decide how to perform this test?

A depth of penetration formula using *resistivity*, *frequency*, and *permeability* can be expressed as shown below. Initially, this may look very complex, but we can simplify it to a **three-step process**. For nonferromagnetic materials the term μrel is ignored.

Step 1 The value is assumed to be 1.
Step 2 Any number times 1 is the same number.
Step 3 Therefore, frequency (f) times the relative permeability (μrel) simplifies to just the frequency (f) <u>in Hertz.</u>

$$\delta = K\sqrt{\frac{\rho}{f\mu_{rel}}}$$

Where: (delta) δ = Depth of penetration in inches
(Constant) K = 1.98
(rho) ρ = Resistivity in $\mu\Omega$cm
f = Frequency in Hertz
μ_{rel} = 1 for non-ferromagnetic materials

The "K" factor becomes 1.98 if we are going to be working in inches. Let's just call it 2. So our formula for the SDP of nonferromagnetic material simplifies to:

$$SDP = 2\sqrt{\frac{\rho}{f}}$$

What we now have is a simple math problem with two variables. All we need is the material resistivity (or an IACS table of conductivity values so that we can convert) and the test frequency we will be using. With these two numbers we can calculate our SDP. Our prime variable is *frequency* as eddy current inspectors. By adjusting the frequency, we can be selectively responsive to test object variables.

Knowledge of our "standard" and "effective" depths of penetration (SDP and EDP) will tell us how our eddy currents are distributed in the material and how strong they are in our area of interest. This information may provide us clues as to what output signal *amplitudes* we should expect for certain types of localized conductivity changes or how much *phase spread* we should expect for discontinuities at different locations within the material. Let's try it.

- We have two aluminum plates 0.25 inch (6 mm) thick. It is suspected that the material interface may be undergoing an active corrosion mechanism. We do not have a specific procedure to follow and we want to know which frequency to use to give sensitivity at the interface 0.25 inch (6 mm) below the surface).

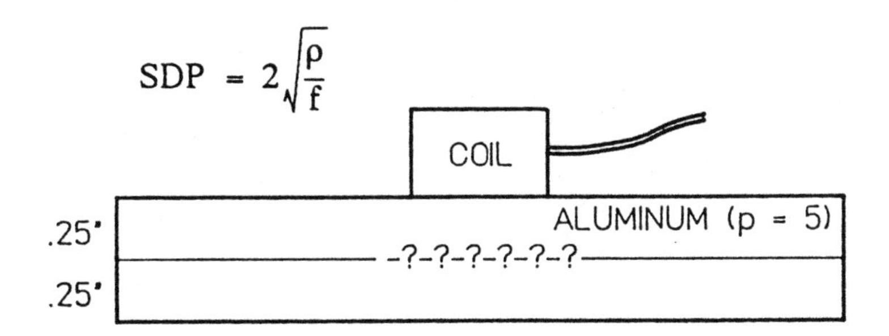

Lets use the 1 kHz range as a good starting point. Our SDP formula, would give us:

$$SDP = 2\sqrt{\frac{\rho}{f}} = 2\sqrt{\frac{5}{1000}} = 2\sqrt{0.005} = 2(0.07) = 0.14''$$

If we assume that the effective depth of penetration (EDP) is three times the value then: EDP = 3 (0.14) = 0.42 inch. This has two advantages:

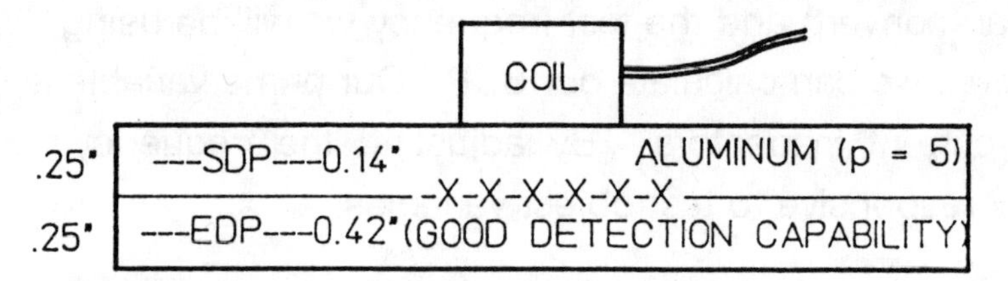

- The **first advantage** is that we will reach the interface of the two plates (0.25 inch) which will occur before we reach our second depth of penetration limit (2 x SDP = 0.28 inch). This means that we should have about 15 percent of our original field strength remaining available to us to attempt to detect conductivity (material) changes at the interface between the two plates.

- The **second advantage** is that our EDP (3 x SDP = 0.42 inch) is less than the total plate thickness (0.50 inch). By maintaining this relationship, we can assure that our inspection will not be affected by anything occurring outside of our area of interest.

Let's say that we decide to go ahead and perform this inspection with our 1 kHz inspection frequency. Our eddy current field intensity at the interface (15 percent ?) may not really be adequate to detect small material changes at the interface. To improve our sensitivity, we would need to get a greater percentage of our available energy deeper into the material. One way to improve this sensitivity at 0.25 inch into the test piece would be to decrease our frequency. This would lead to greater SDP and EDP thicknesses.

However, keep in mind that if our EDP gets much larger (greater than 0.5 inch), then we are going to become sensitive to changes beyond our back wall. If we allow that to happen, then we would have to fully understand what possible conditions in that area might be and how they might affect our test results.

SUMMARY

We have discussed those factors that will affect how we perform our examination of different materials. The major controlling factors are:

MATERIAL PROPERTIES: Conductivity / σ (or Resistivity / ρ)
Permeability / μ
Geometry (Dimensional Parameters)

If we can understand the relationship of these quantities, then it will be possible for us to select the proper inspection frequencies.

- As resistivity increases, conductivity decreases.
- As conductivity increases, resistivity decreases.

By understanding how frequency affects the distribution of our eddy currents, we will be better able to know where our inspection is most sensitive during our testing.

$$SDP = 2\sqrt{\frac{\rho}{f}}$$

$1 \times SDP = 1/e = 37\%$ (where $e = 2.718$)
$2 \times SDP = 1/e^2 = 14\%$
$3 \times SDP = 1/e^3 = 5\% = EDP$

By knowing the areas of highest test sensitivity, we should be able to assure ourselves and our customers that we are providing them the best possible inspection.

CHAPTER 3

EDDY CURRENT INSPECTION COILS

As you might guess, eddy current testing inspection coils are designed for a very wide range of applications. In looking at some typical coils, it can be seen that they can be separated into **three basic types** related to their physical structure and the type of testing that they perform.

- Probe Coils
- Bobbin Coils
- Encircling Coils

Probe Coils may sometimes be addressed as surface coils, flat coils or pancake coils. These are common terms used to describe the same test coil type. The probe coils shown in Figure 3-1 provide a convenient method of examining the surface of a test object.

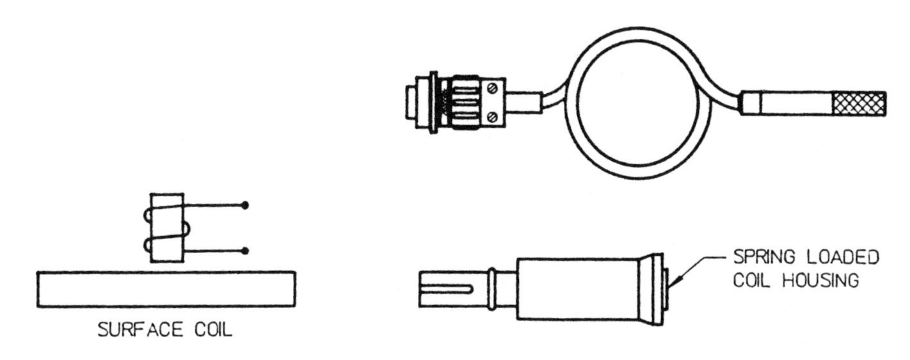

Figure 3-1
Typical Surface Scanning Probe Coil

Figure 3-1 illustrates a typical probe coil used for surface scanning. There are several classical features to this probe.

- Spring mechanism for minimizing *lift-off* (maintaining good surface contact).

- Wire slightly recessed with a wear-resistant plate to protect the coil wires. The coil would be secured with a layer of epoxy. This not only secures the windings and leads in place but also provides some wear resistance. This will prevent electrical shorts from occurring and extend probe life.

- A nonconductive (instead of metal) housing prevents heat transfer to the coil from the inspector's hand or other external sources.

- Modular (two-part) design incorporating both a probe head and a reinforced cable. This way, when you try to pull the tester along behind you by the cable, all you have to replace is the cable you just tore up—not the entire probe.

As shown in Figures 3-2 and 3-3, probe coils and probe coil forms can be shaped to fit particular geometries to solve complex inspection problems.

As an example, probe coils fabricated in a pencil shape (pencil probes) might be used to inspect an interior surface through a hand hole or an interior radius of a turbine blade.

Pancake coils can be mounted on a holding device and used to inspect the inside radius of a tube or bolt hole. These could be addressed as "inside coils," "pancake coils," or "surface probes" since the test arrangement fits all these categories.

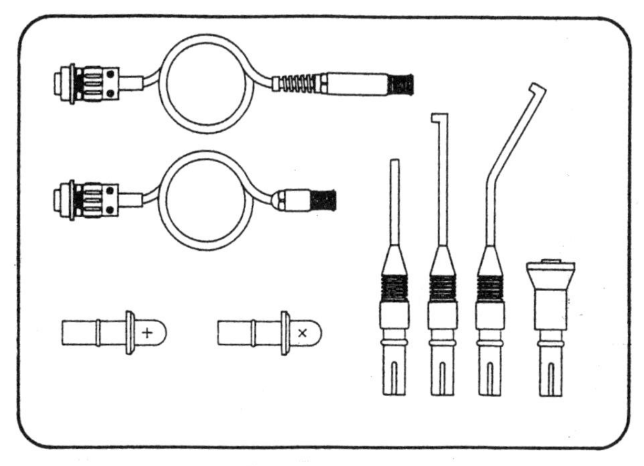

Figure 3-2
Selection of Specialized Surface Probes

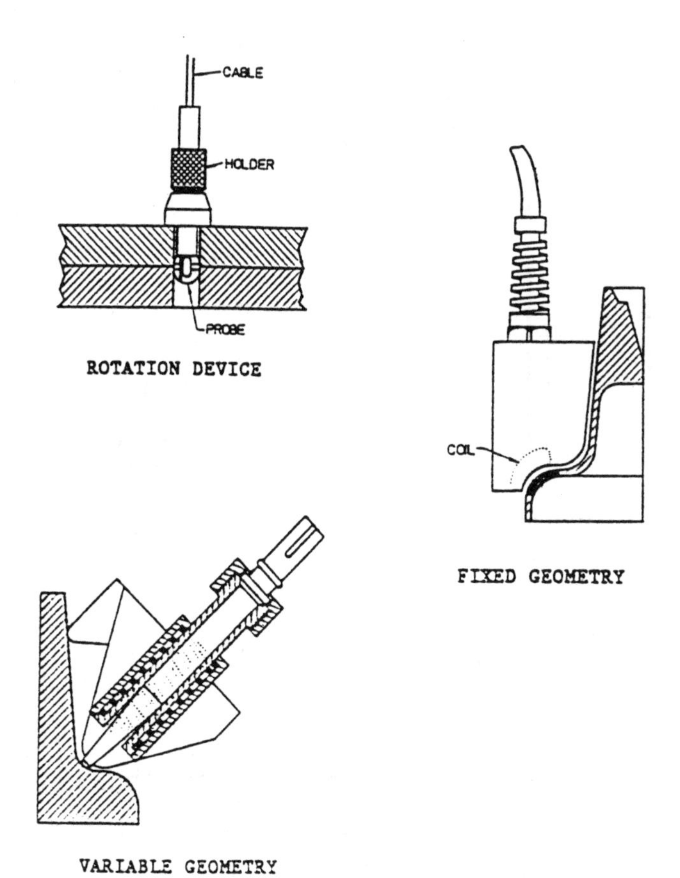

Figure 3-3
Specialized Holding Fixtures

Probe coils may be used where high resolution is required by adding coil shielding. Shielding can be provided by surrounding the coil in either ferritic or conductive materials. As shown in Figure 3-4, the use of "cup cores" or "mu metal" are common. These materials limit the magnetic field extension outside of the probe holder's geometry.

The advantage to using shielding is that we will restrict the horizontal extension of the primary magnetic field. By doing this, we may be able to perform an inspection near the edge of an inspection sample. By restricting our field of view, we can improve our sensitivity to flaws in the area of interest and not be affected by unwanted signal sources close to the inspection zone. A slight disadvantage is that there is also *some decrease in the depth of penetration*. One could not achieve the same SDP and flaw sensitivities with the same coil both shielded and unshielded. The ability to restrict our field of view normally leads to an improved signal-to-noise ratio in the area of interest.

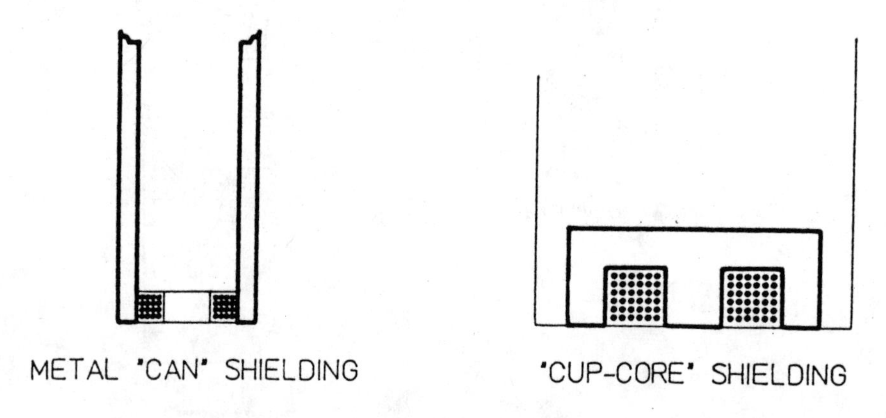

Figure 3-4
Coil Shielding to Limit Magnetic Field

When using a high-resolution probe coil, the test object surface must be carefully scanned to assure complete inspection coverage. This careful scanning is very time consuming. For this reason, probe coil inspections of large test objects are usually limited to critical areas.

As shown in Figure 3-5, probe coils are used extensively in aircraft inspection for crack detection near fasteners and fastener holes. In the case of cracks actually inside the fastener holes (bolt holes, rivet holes), the probe coil is spinning while being withdrawn at a uniform rate. This provides a helical scan of the hole using a "spinning probe" technique.

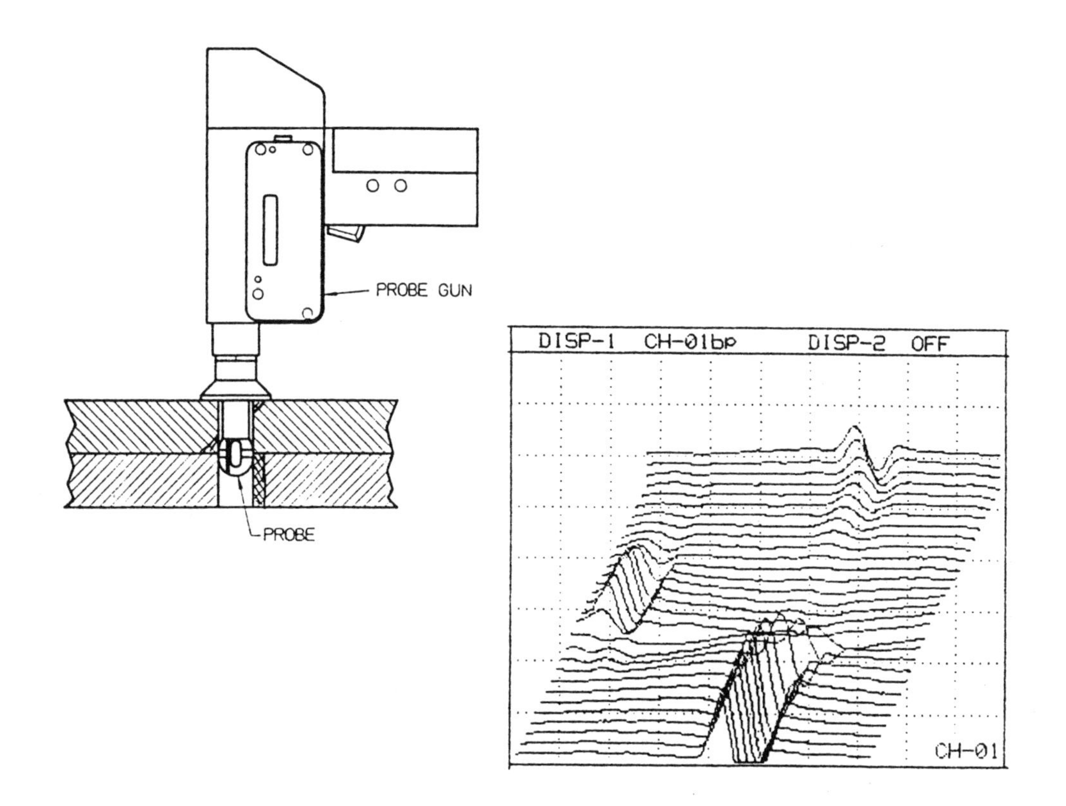

Figure 3-5
Indexing Probe Gun and Three-Dimensional Flaw Plot

If we could see eddy currents in a material, they would look like the mirror image of the electrical current flowing in the wire of our inspection coil. The direction of the windings in a typical probe coil is parallel to the surface we are inspecting. The eddy currents will also be flowing parallel to the surface. The shape and orientation of the winding dictates what the eddy current flow path is going to look like (Figure 3-6).

Looking down at the surface of a flat plate from an angle would show us more information about these "swirling" eddy currents as shown in Figure 3-7. They form circular patterns in the material that look like tornados. Their dimensions are controlled by the size and type of coil, the frequency and amperage being supplied to the coil, and the electrical properties of the material being inspected.

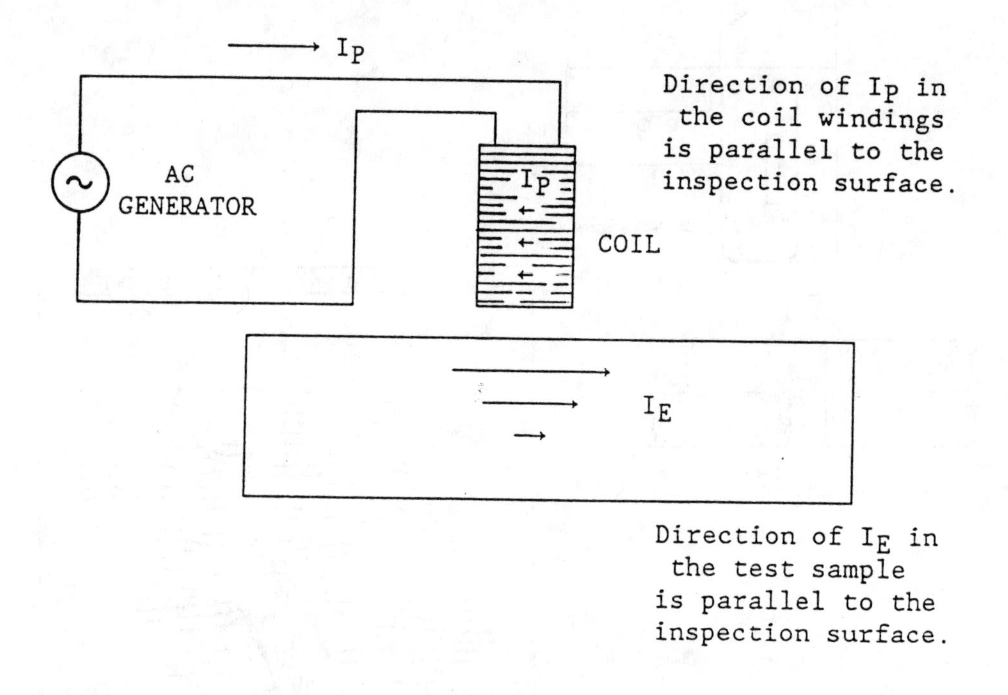

Figure 3-6
I_P Compared to I_E

As we discussed in the last chapter, we are going to have a greater intensity (or distribution) of eddy currents at the surface than are going to be found deeper in the material.

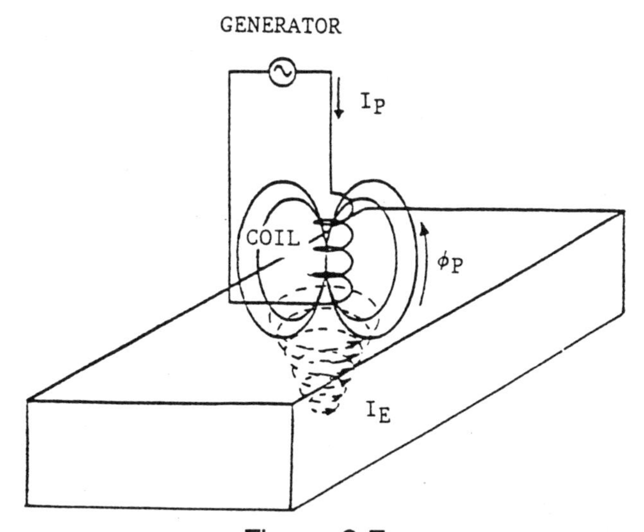

Figure 3-7
Eddy Current Flow Pattern Using Probe Coils

Encircling Coils are the second major category of probes based on their mechanical features. These are sometimes known as OD coils, or "feed-through" coils, because the coil surrounds the test object. Figure 3-8 illustrates a typical encircling coil application with a circumferential crack.

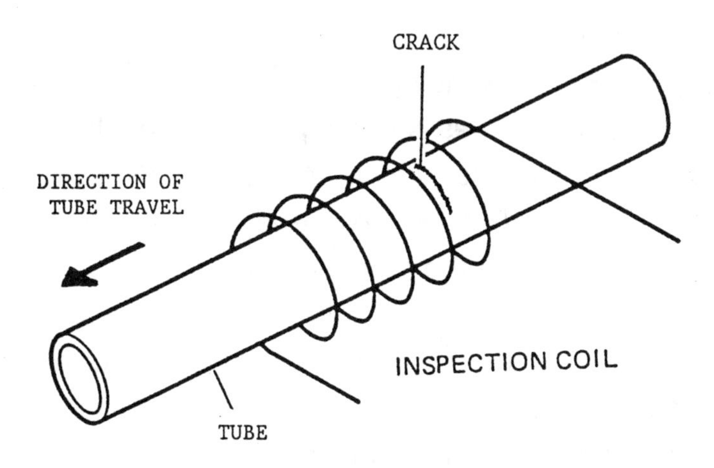

Figure 3-8
Encircling Coil Inspection of Round Stock

An important element to consider prior to deciding which eddy current test method and probes to apply is that we need some knowledge of the "typical" or expected flaw sizes and orientations in the material to be tested. In Figure 3-9 we see a crack orientation (longitudinal) that would be much easier to see when using the encircling coil technique.

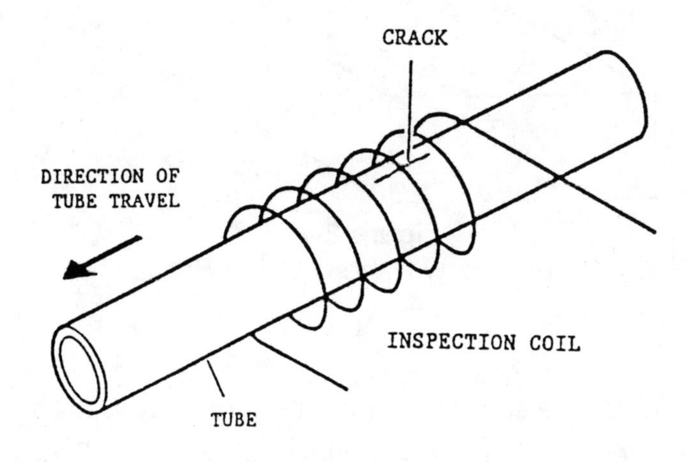

Figure 3-9
Encircling Coil Inspection of Round Stock

As shown in Figure 3-9, the windings go around the tube. That means that our primary current in the wires is flowing around the tube circumference. We stated earlier that eddy currents always flow in the opposite direction as the primary current; therefore, our eddy currents must be flowing around the circumference of the tube as well.

Shown in Figure 3-10 is the distribution of the eddy current field on the surface near the coil and again at the end of the rod showing a possible distribution based on the depth of penetration.

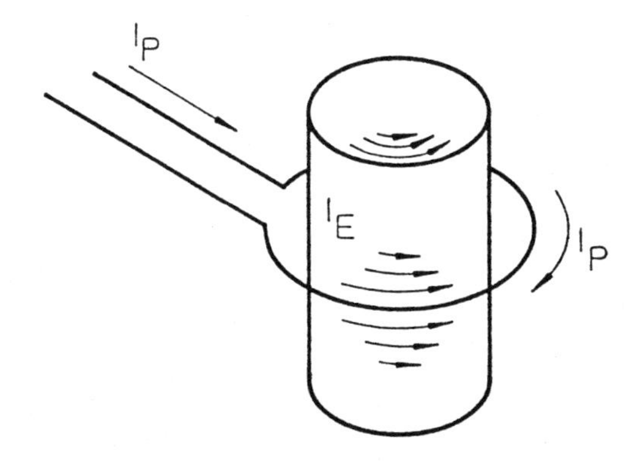

Figure 3-10
Encircling Coil Inspection of Round Stock

Encircling coils are primarily used to inspect tubular and bar-shaped products. The tube or bar is fed through the coil (feed-through) at a relatively high speed. The cross section of the test coil is simultaneously interrogated. For this reason, circumferential discontinuities are difficult to detect with an encircling coil. The volume of material examined at one time is greater using an encircling coil than a probe coil; therefore, the relative sensitivity is lower for an encircling coil. When using an encircling coil, it is important to keep the test object centered in the coil, as illustrated in Figure 3-11.

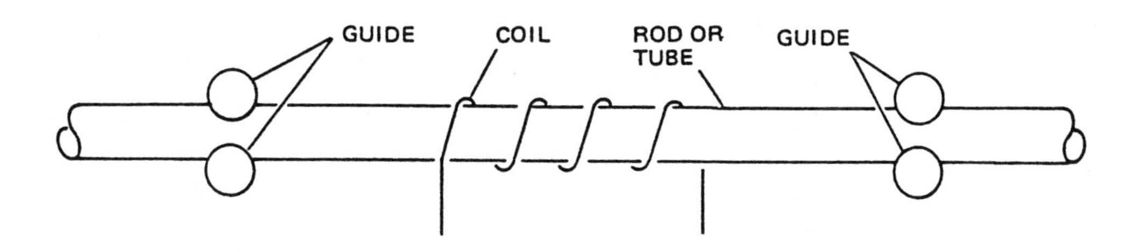

Figure 3-11
Encircling Coil in an On-Line System

If the test object is not centered, as shown in Figure 3-12, a uniform discontinuity response is difficult to obtain. It is common practice to run the calibration standard several times, each time indexing the artificial discontinuities to a new circumferential location in the coil. This procedure is used to ensure proper response and proper centering.

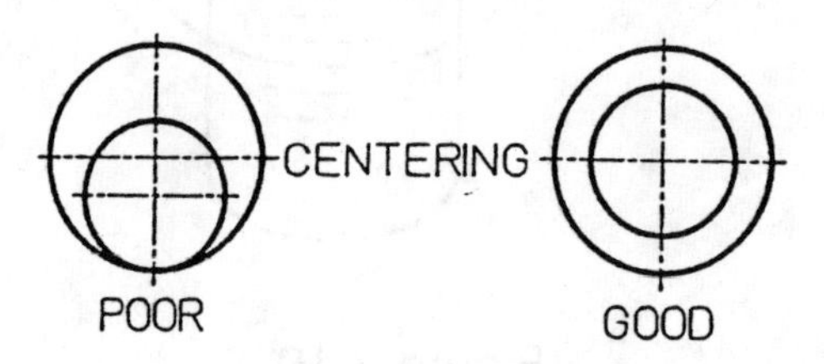

Figure 3-12
Proper Centering of Test Sample

Bobbin Coils are sometimes called ID coils and/or inside probes. These names describe coils used to inspect from the inside diameter or bore of a tubular test object. Bobbin coils are inserted and withdrawn from the tube ID by long, semiflexible shafts or simply blown in with air and retrieved with an attached pull cable. These mechanisms will be described later in the text. Bobbin coil information follows the same basic rules stated for encircling coils. Figure 3-13 illustrates a typical bobbin coil.

Probe coils, encircling coils, and bobbin coils can be additionally classified. These additional classifications are determined by how the coils are electrically connected. The three coil categories are absolute, differential, and hybrid. Figure 3-14 shows various types of absolute and differential coil arrangements.

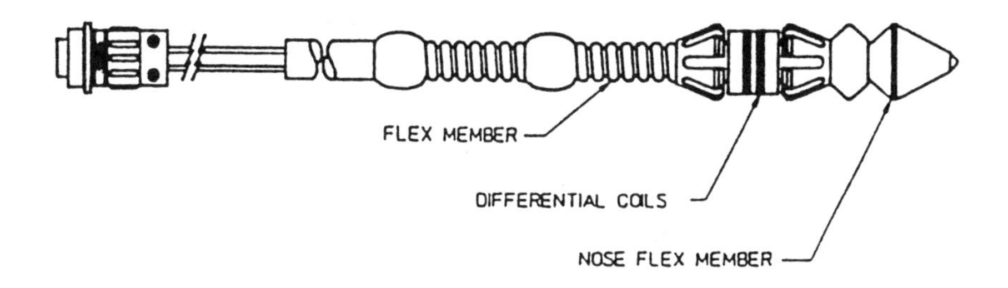

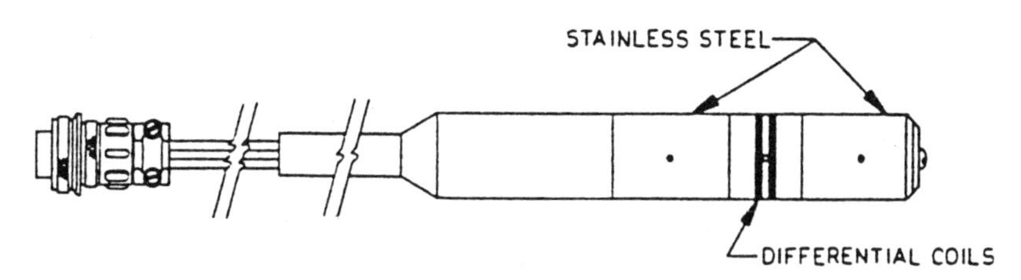

Figure 3-13
Typical Bobbin Coil

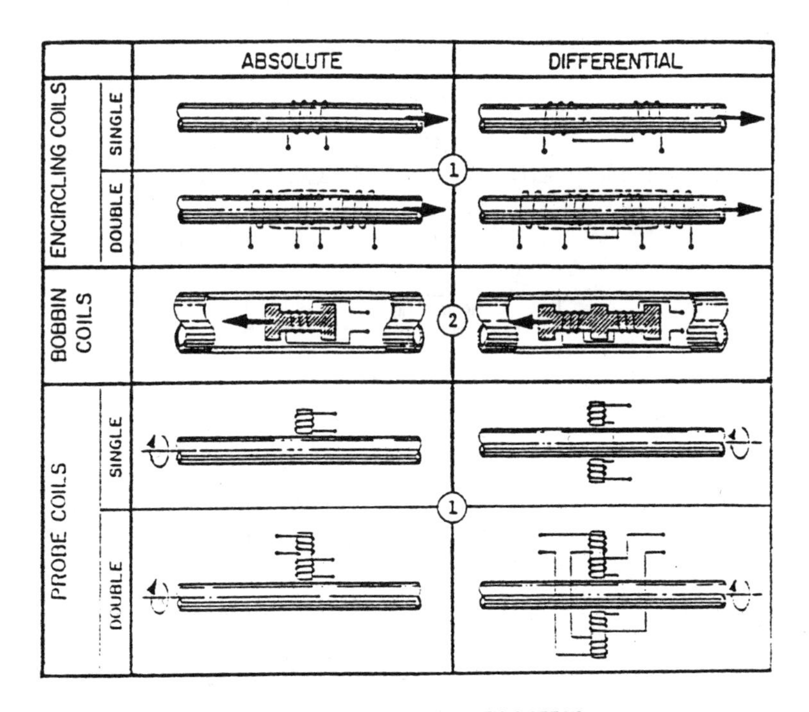

1. TEST PIECE IN MOTION
2. TEST COIL IN MOTION

Figure 3-14
Coil Configurations

Absolute Coils

An **absolute coil**, as shown in Figure 3-14, can be defined as making a measurement *without direct reference or comparison to a standard* as the measurement is being made. Some applications for absolute coil systems would be measurements of conductivity, permeability, dimensions and hardness as shown in Figures 3-15 and 3-16.

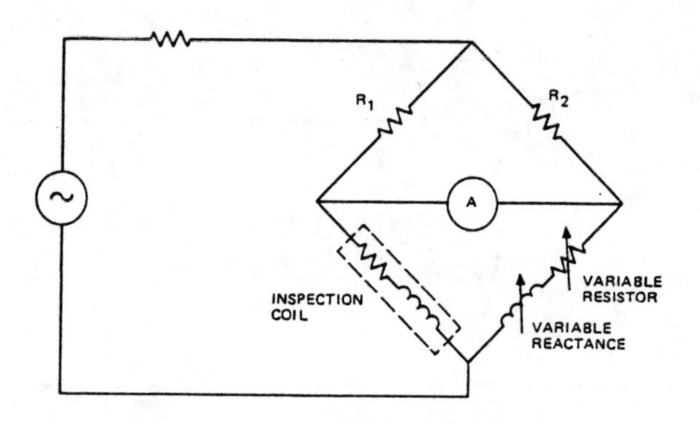

Figure 3-15
Circuit Diagram for Absolute Coil Applications

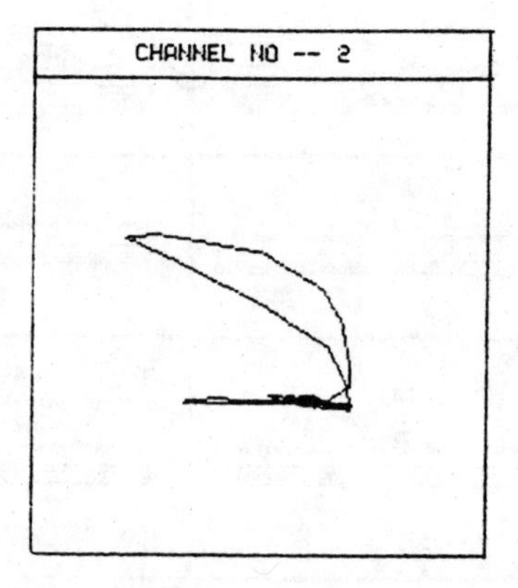

Figure 3-16
X-Y Screen Presentation of Absolute Signal Response

Differential Coils

Differential coils consist of two or more coils electrically connected to oppose each other. Differential coils can be categorized into two types.

The first type is the **"self-comparison differential**." The self-comparison differential coil compares one area of a test object to another area on the same test object as shown in Figure 3-17. A common use is two coils connected opposing each other in the balance circuit. If both coils are affected by identical test object conditions, the net output is "0," or no signal change.

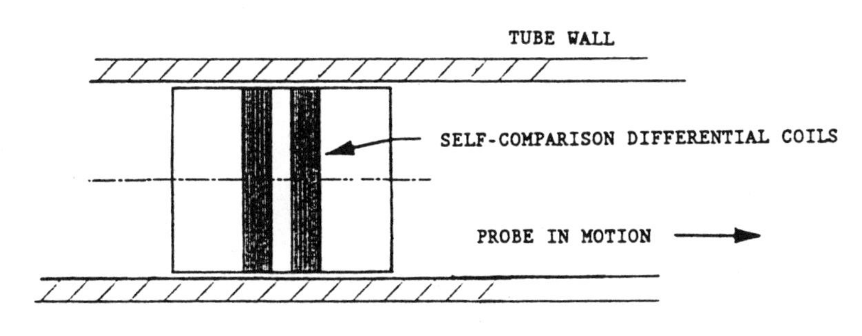

Figure 3-17

Self-Comparison Differential Coil

The self-comparison arrangement is insensitive to test object variables that occur gradually or have what might be called a "slow rise time." Variables such as slowly changing wall thickness, diameter, or conductivity are effectively discriminated against with the self-comparison differential coil.

Only when a different condition is presented to one or the other test coils is an output signal generated. The coils usually being mechanically and electrically similar allows the arrangement to be very stable to temperature changes. Short discontinuities or events with a "fast rise time," such as cracks, pits, or other localized discontinuities with abrupt boundaries, can be readily detected using the self-comparison differential coil. One possible signal pattern from this type of approach is seen in Figure 3-18.

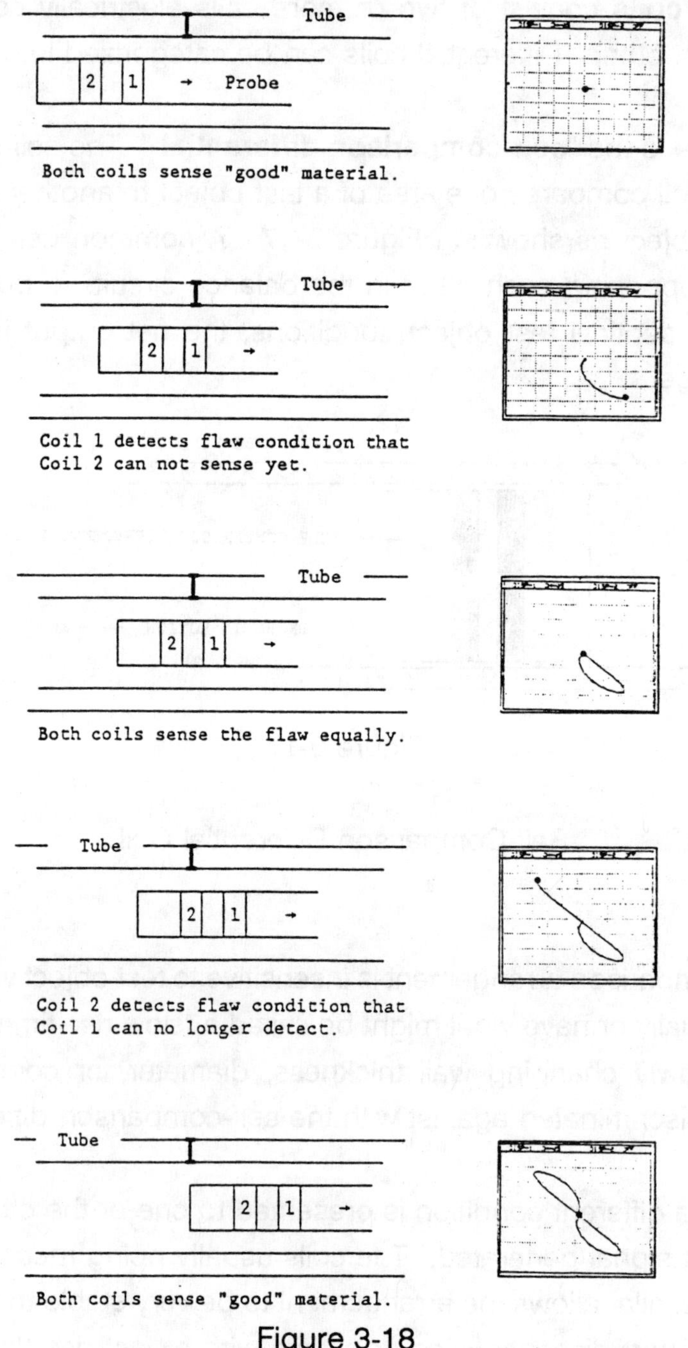

Figure 3-18
Self-Comparison Differential Responses

The second differential arrangement is the **"external reference differential"** coil. As the name implies, an external reference is used to affect one coil while the other coil is affected by the test object.

External reference differential coils are used to detect differences between a standard object and test objects. It is particularly useful for comparative conductivity, permeability and dimensional measurements.

It becomes obvious in Figure 3-19 that it is imperative to balance the system with one coil affected by the standard object and the other coil affected by an acceptable test object. The external reference differential coil system is **sensitive to all measurable differences** between the standard object and test object. For this reason, it is often necessary to provide additional discrimination to separate and define variables present in the test object.

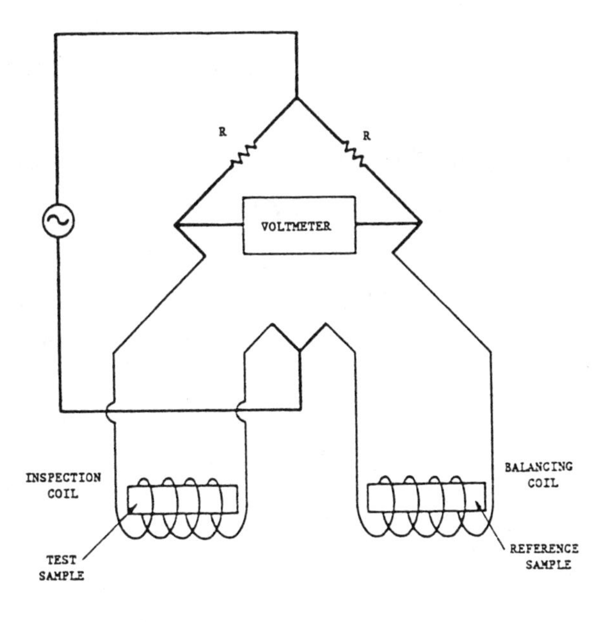

Figure 3-19
External Reference Differential Coils

Hybrid Coils

Hybrid coils may or may not be the same size and are not necessarily adjacent to each other. Common types of the hybrid coils are driver/pickup, through-transmission, or primary-secondary coil assemblies. Figure 3-20 shows a typical hybrid arrangement.

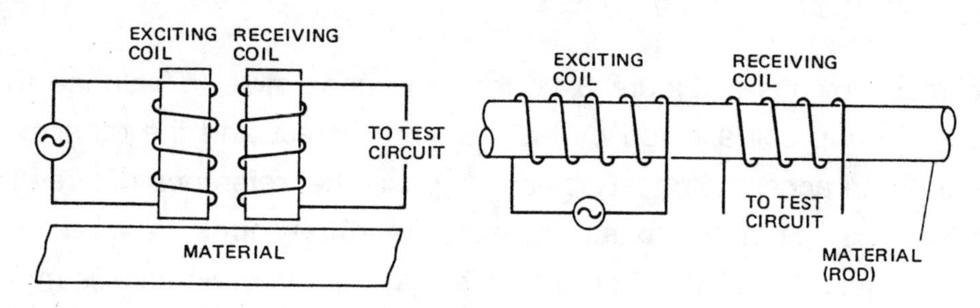

Figure 3-20
Hybrid Coil Arrangements

A simple hybrid coil consists of an excitation coil and a receiving (sensing) coil. The primary or "driver" coil is actually the only one of the two that has an AC signal being supplied to it by the tester. The secondary or "pickup" coil signal shifts are being induced by variations in I_E and ϕ_E.

In the through-transmission coils, the excitation coil is on one side of the test object and the sensing coil is on the other as shown in Figure 3-21.

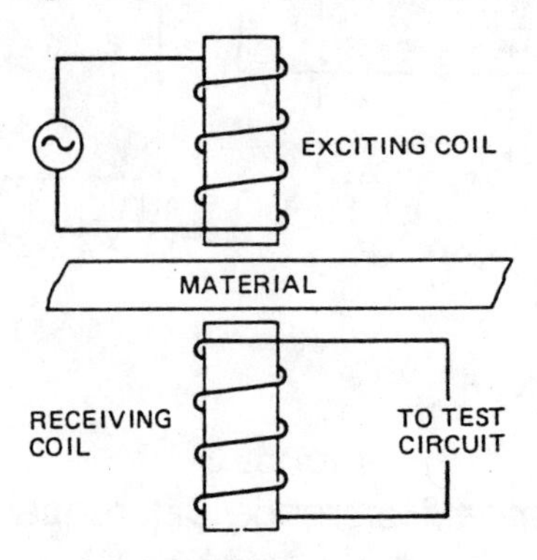

Figure 3-21
Through-Transmission Coil Arrangement

When using driver/pickup coil arrangements, the voltage developed in the sensing or pickup coils is a function of:

- the current magnitude and frequency applied to the excitation coil.
- coil parameters of the exciting and sensing coils (size, shape, and orientation of the test sample and to each other).
- test object characteristics.

SUMMARY

The three main coil groups based on their structure are:

- Probe (or surface) coils
- Encircling (or outside) coils
- Bobbin (or inside) coils

The three basic modes of operation are:

- Absolute
- Differential
- Hybrid

The choice of which type of probe to use and how each will be energized and sampled is a key element in determining the quality and efficiency of the examination that we are performing.

CHAPTER 4

EDDY CURRENT TEST CIRCUITS

In Chapter 1 we discussed the principles of electromagnetic induction. In Chapter 2 we discussed the factors of conductivity, dimensions, and permeability and their effect on the magnetic field of the test coil. In this chapter we are going to show you how the presence of these effects are actually detected.

You will recall that when an alternating current is applied to a coil the amount of current that flows through the coil is determined by the impedance of the coil. You will also recall that impedance is the result of the resistance and the inductive reactance of the coil. In the simple test circuit shown schematically in Figure 4-1, we are showing the inspection coil which consists of a resistance (R) and an inductive reactance (X_L) connected to a source of alternating current. We have also placed an ammeter in the circuit to measure the amount of current flowing through the coil.

If we increase the value of the resistance, the current flowing through the ammeter will decrease. Since there is more resistance to current flow, less current will flow, and the ammeter reading will decrease. Increasing either the inductive reactance or resistance will increase the impedance of the circuit, thus causing the current to decrease.

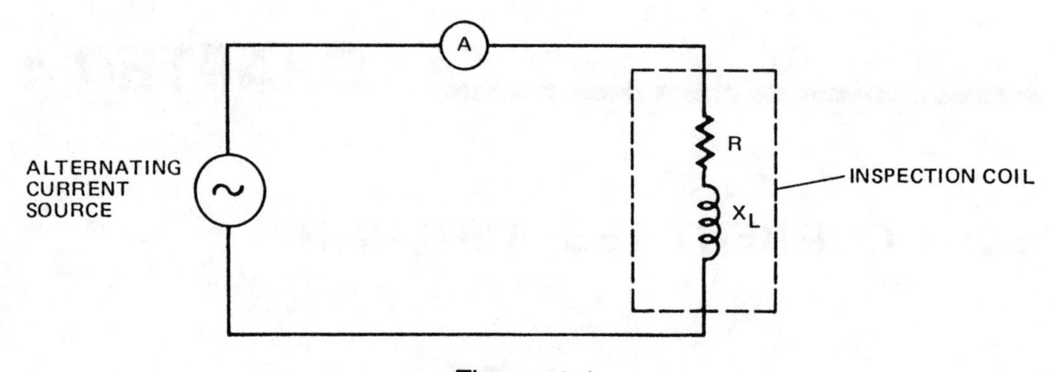

Figure 4-1
Simple AC Test Circuit Schematic

Now let's look at the circuit again from a different viewpoint as shown in Figure 4-2. Suppose that we have read the ammeter with the coil held in the air and obtained a reading of 10 amperes. We then place the coil on top of a piece of copper and the meter reads 9.5 amperes. Since nothing else in the circuit changed, the only thing that could cause a change in the meter reading would be that the impedance of the coil changed somehow.

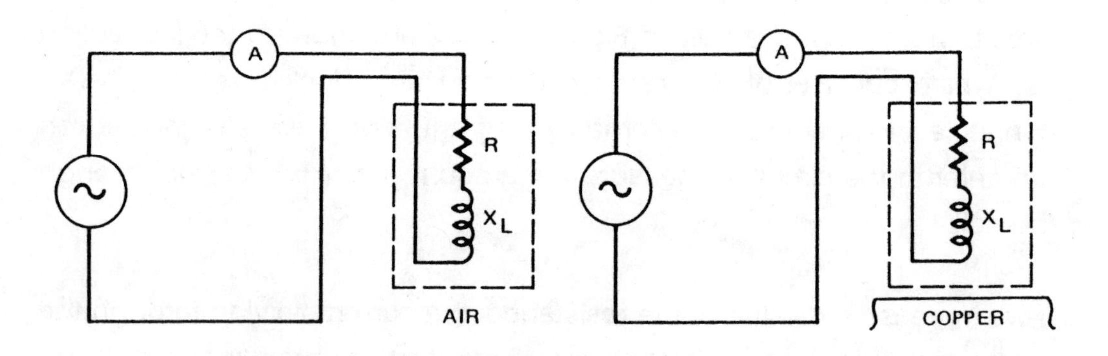

Figure 4-2
Measurable System Change from Air to Test Sample

This is the link we needed. In Chapter 2 we discussed the ways in which the coil's magnetic field was affected by a change in conductivity. Now we have shown that a change in conductivity of the material under the inspection coil (from air at 0 percent IACS to copper at 100 percent IACS) has caused a change in a meter reading. From this we can deduce that a change in the coil's magnetic field causes a change in the impedance of the coil. And now we know that this change will cause a change in the reading on the meter; something that we can see and record.

The block diagram in Figure 4-3 shows the basics of what is occurring in the test we just described. Everything that is happening in the material is shown below the dashed line. Everything that is happening in the test circuit is shown above the dashed line. The box in the center of the illustration shows the deduction that we just made as a result of the test we just described.

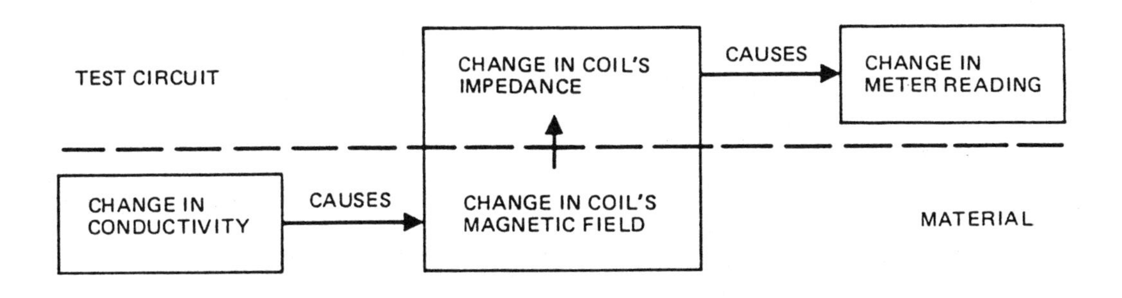

Figure 4-3
Block Diagram Showing the System Response Logic

The change in the coil's magnetic field causes a change in the coil's impedance. This is the link between the material and the test circuit.

Notice that we call the coil, the meter, and the AC power supply a "test circuit." We do this because the test circuits used in eddy current testing become very complex as we go along and include more than the components that we have shown here. While the circuit shown works as advertized, it is not very sensitive to the minute material changes we will be looking for. For example, we obtained only a half-amp change in meter reading for a 100 percent change in conductivity. This circuit would not be very sensitive to a change in conductivity caused by a small crack. A more sensitive test circuit can be obtained with the basic bridge circuit.

As shown in Figure 4-4, the basic bridge circuit consists of two balanced resistors (R_1 and R_2), the inspection coil, a balancing impedance, and an ammeter. These units are connected in a bridge format; i.e., a resistor and the inspection coil in one leg of the bridge, a resistor and a balancing impedance in the other leg of the bridge, and an ammeter across the two legs. The AC source is connected across the bridge.

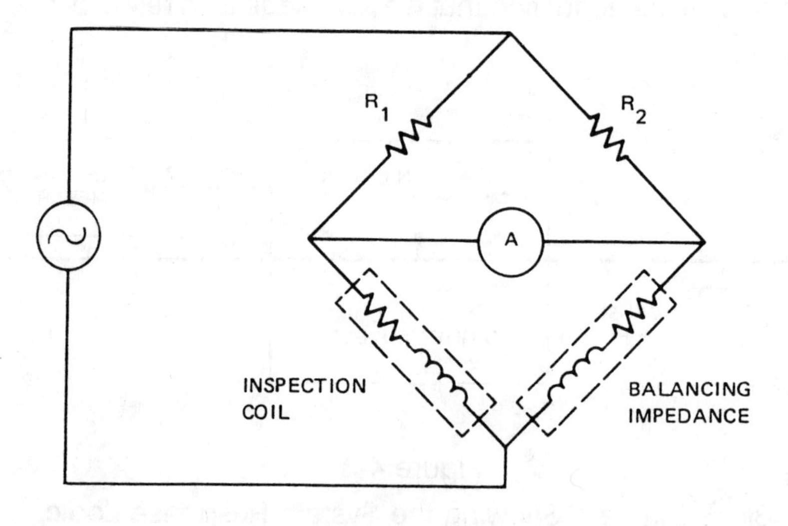

Figure 4-4
Basic Bridge Circuit

When the bridge is in balance (resistors and impedances equal) the meter will read zero amps. When the inspection coil is placed on or near a test sample, there is a change in impedance in that leg of the bridge, the bridge becomes unbalanced, and the ammeter will indicate a current that is proportional to the imbalance. The bridge circuit shown in Figure 4-4 is far more sensitive to impedance changes than the basic test circuit. A reading on the meter *only* means that the bridge is unbalanced. It means nothing else, ever. It is up to the test technician to know why it is unbalanced and what the unbalance means with respect to the test he is conducting.

Some simple eddy current instruments are shown in Figure 4-5. In all cases, since only the voltage change or magnitude is monitored, these systems can all be grouped as *impedance magnitude types.*

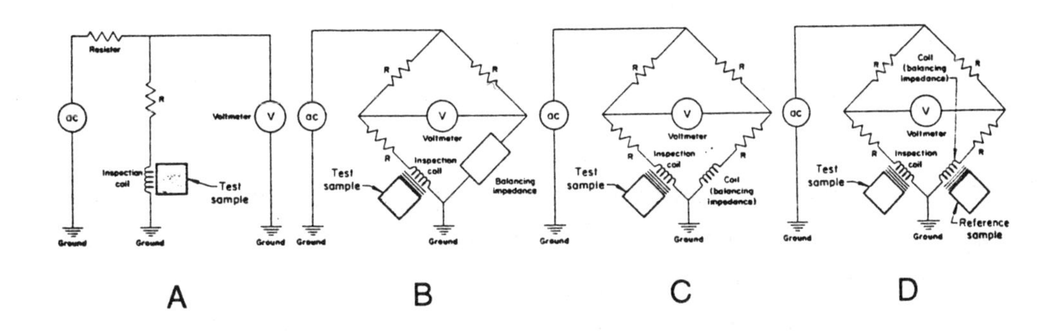

Figure 4-5
Four Types of Simple Eddy Current Instruments

- Figure 4-5 (A) shows the voltage across the inspection coil being monitored by a simple voltmeter.

- Figure 4-5 (B) shows an impedance bridge circuit. Notice that the balancing impedance (Z) is "fixed."

- Figure 4-5 (C) has variable resistors and inductors to allow a wider balance range for a larger number of applications.

- Figure 4-5 (D) illustrates a balance coil affected by a reference sample. This is commonly used in external reference differential coil tests.

We mentioned earlier that the electrical "changes" that may indicate some material change are extremely small. The circuitry required to detect, measure, and display these changes can be simplified into the block diagram shown in Figure 4-6.

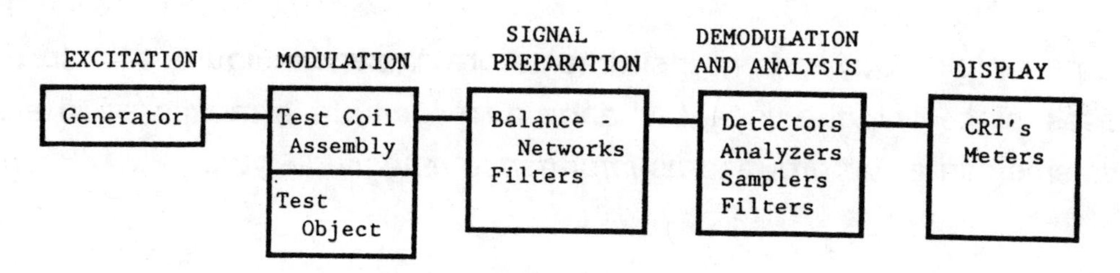

Figure 4-6
Internal Functions of an Eddy Current Test System

Most eddy current instrumentation is categorized by its final output or display mode. There are basic requirements common to all types of eddy current instrumentation. **Five different components** are usually required to produce a viable eddy current instrument. These functions are:

- Excitation - The generator provides power to the coil.

- Modulation - Occurs at the primary and secondary magnetic field interface.

- Signal Preparation - Balance networks are used to "null out" steady-value alternating current signals. Amplifiers and filters are also part of this section to improve signal-to-noise ratio and raise the signal levels for further processing and display.

- Demodulation and Analysis - In this section, we may find various detectors, analyzers, discriminators, filters and/or sampling circuits. Detectors can be a simple amplitude type or a more sophisticated phase-amplitude or coherent type.

- <u>Signal Display</u> - The signal display section is the key link between the test equipment and its intended purpose. The signal can be displayed many different ways. Common displays include meters, cathode-ray tube oscilloscopes (CRTs), LCD-type displays, and computer screens. Any or all of these may incorporate visual and/or audible alarms.

An optional **sixth component**, as shown in Figure 4-7, could be shown, depending on the complexity of the examination. If a Go-NoGo criteria is in place, the system may be linked to various relays or test object handling equipment.

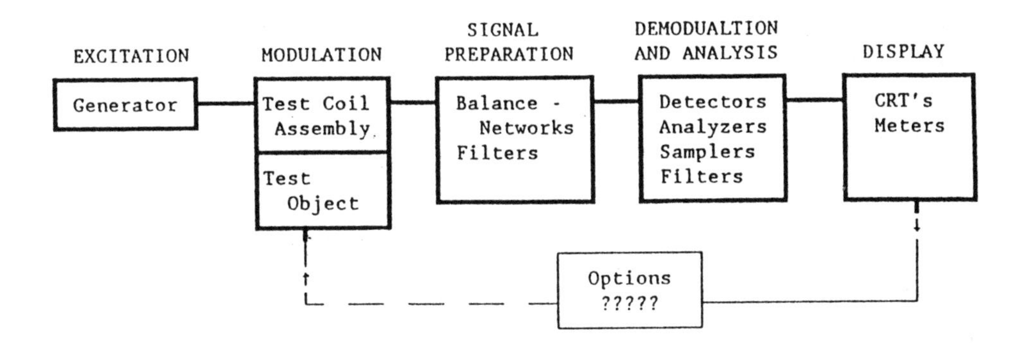

Figure 4-7
Options: Feedback Controls or Recording Devices

- Audio alarms can be used in much the same manner. Usually the audio alarm indicates only the abnormal condition. Alarm lights and audio alarms are commonly incorporated in eddy current test equipment. The indicator light and audio alarm give only **qualitative** information about the item, whether a condition is present or not. The degree of the condition cannot normally be determined with these devices. Indicator lights and audio alarms are relatively inexpensive and can be interpreted by semi-skilled personnel.

- If the examination requires data storage, then there could be any of a wide range of analog or digital recorders used. These would provide long-term retrieval of the data for later review and analysis.

- Strip-chart recordings are common in testing tubing or nuclear fuel rods where the discontinuity location along the length of the rod or tube is critical. The strip-chart length is indexed to time or distance and the pen response indicates either normal or abnormal conditions.

Summary

In previous chapters we have dealt with the factors of inducing an eddy current flow in a material. What we can surmise from these discussions is that:

- The eddy currents that are actually induced in the secondary materials are quite small when compared to the strength of the primary current in the coil (Figure 2-1).

- These eddy currents are going to be affected by material variations in the test sample. They are going to be affected by disruptions to the localized conductivity (discontinuities) in the material. These material changes will affect the way the eddy currents move in the material.

- These changes in the eddy current flow path are reflected back through the coil when the primary and secondary magnetic flux field strengths change and alter the circuit impedance. If we hope to detect when these eddy currents are forced to move out of their normal flow patterns, then we are going to have to rely on some pretty sophisticated measuring circuitry.

- Bridge circuits, as well as the associated balance, detection, and display circuitry in modern eddy current test equipment, allow us the possibility of detecting and characterizing smaller and smaller variations in the eddy current examination.

CHAPTER 5

EDDY CURRENT RESPONSE CHARACTERISTICS

Now that you have a general idea of how eddy current testing works, let's get down to the nitty gritty. In this chapter, we are going to discuss some of the factors that affect our eddy currents during a test. These factors include the orientation of the eddy currents with respect to the coil producing them as well as how changes in the material or the inspection geometry will affect the test.

We have discussed briefly the orientation of the eddy currents with respect to the coil. We said that eddy currents flow in circular paths that are parallel to the windings of the inspection coil.

The purpose of the inspection determines, to a considerable extent, the type of coil to be used. When inspecting for discontinuities, it is essential that the flow of eddy currents be as nearly perpendicular to the discontinuity as possible to obtain the maximum response from the discontinuity. If the eddy current flow is parallel to the discontinuity, there will be little or no distortion of the currents and therefore little or no detection capability. In Chapter 2 we determined that three kinds of material variations might affect our eddy current examination: *conductivity, permeability,* and *part geometry* or *dimensional characteristics.*

We discussed how material permeability may affect our eddy current exam. Depending on the strength of the inherent magnetic domains in a material, we may have to attempt to "saturate" the material. If we could use either direct (DC) current or permanent magnets to overcome the permeability factor, then we might be able to perform a more reliable test.

For the range of materials that we would most often be testing with eddy current, it is more likely that the equipment would be affected by either conductivity or dimensional variations. A change in conductivity can be detected by eddy current testing. Since various materials do have different conductivities, we can **sort conductive materials** by measuring the conductivity of each piece.

But—there are other factors which can change the conductivity of a sample of given material. While the inherent conductivity of a material is always the same, there are internal factors that can cause what appears to be a change in the inherent conductivity. The following **five major factors** can change conductivity.

- A combination of two or more materials to form an **alloy**
- Changes in the **hardness** of the material due to heat treatment
- Changes in the **temperature** of the material
- Residual **stresses** in the material
- The presence of a thin coating of **cladding** of another conductive material

Let's discuss the factors that change the conductivity of a material, one at a time.

- **Alloys**

  Alloys are combinations of other metals and/or chemical elements with a base metal to form what is essentially a new and different material. Its properties, including that of conductivity, are different from the base material. The new material, so long as it has the same proportional amounts of its elements, will always have the same conductivity. A variation of any of the alloy elements will cause the conductivity to change. This change in conductivity is detectable with eddy current instruments, provided that the change in conductivity is large enough to be detected. The purity of the alloy can be tested in this manner when alloying elements increase or decrease the conductivity of the alloy.

An alloy *does not have the same conductivity as that of one of its base metals*. An alloy of different materials always has different properties (hardness, tensile strength, malleability, etc.) than the parent materials. Conductivity is one of those properties. It is this fact that makes eddy current testing most useful in the identification of different materials and their alloys.

- **Material Hardness**

The hardness of some materials (metals) will affect that conductivity. A material that has been heat-treated or that has been overheated in service will also have a change in its conductivity. This change is due to an internal change in the material. If the change in conductivity is great enough (as it might be in age-hardenable aluminum) it is possible to monitor the heat-treat process to assure that the proper hardness is attained.

If a test technician is to attain reliable results from an eddy current test, he must be aware of any possible heat-treat factors that might exist and affect the results he is obtaining. Conversely, if he is getting results that seem out of order, he must be able to confirm that heat-treat or overheating is or is not the cause for the erroneous results.

- **Material Temperature**

  In the same manner, the temperature of the material will affect the conductivity of the material. An increase in the temperature of metals normally results in a decrease in the conductivity. A decrease in the temperature results in an increase in conductivity. The technician must realize this effect exists when he is working with materials at temperatures above or below normal. Fortunately, articles under test seldom have hot spots (the article is often the same temperature all over), so test results are not altered as a result of temperature changes.

- **Internal Stresses**

  Internal stresses, however, do occur in local areas in an article. And these stresses do cause changes in the conductivity of the article at those areas. Here the technician must know the manufacturing processes involved and be able to know when such internal stresses are likely to be present. Localized, high-residual stresses generally decrease the conductivity in that area.

- **Conductive Cladding**

  In Figure 5-1, we show an article which consists of a thin layer of aluminum laid over a thick sheet of copper. The conductivity of the aluminum is 61 percent IACS and the conductivity of the copper is 100 percent IACS.

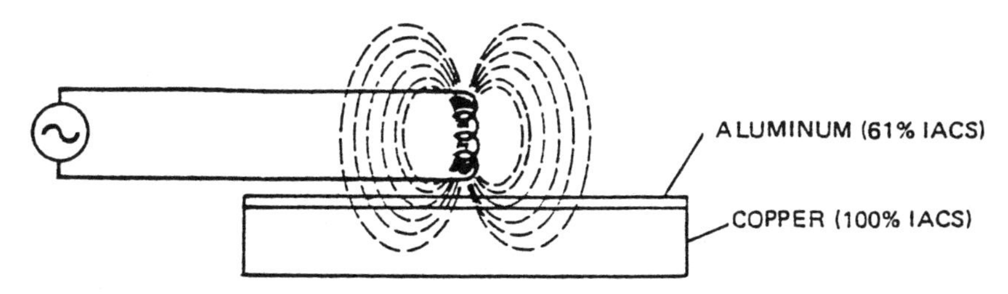

Figure 5-1
Two Material Layers with Different Conductivities

Note that the field of the coil enters both the aluminum and the copper, so both materials will affect the field. If the aluminum had been thick enough so that the field of the coil did not reach the copper, then the conductivity of the article would have been the conductivity of the aluminum only.

The conductivity of this article will be something less than the conductivity of the copper. The aluminum intercepts a small portion of the magnetic field that would otherwise be available for the induction of eddy currents in the copper. The aluminum also has its eddy currents, but they will not be as strong as they would have been for an equal thickness of copper. The conductivity would be something less than 100 percent IACS.

Now, look at Figure 5-2 and suppose that the aluminum cladding was not of equal thickness over the entire surface of the article. Let's assume that the cladding was to be 2-mils (.05 mm) thick, but, due to a problem in the manufacturing process, the cladding over part of the article was only 1-mil (.025 mm) thick. These thin-clad areas can be located by eddy current testing because the conductivity of the article changes as the inspection coil moves from the thicker cladding to the thinner cladding. The conductivity will increase, since there is now more copper and less aluminum in the magnetic field. What we are attempting to show in Figure 5-2 is that *more* of the magnetic field is in the copper when the aluminum cladding is only 1-mil thick than when the aluminum cladding is 2-mils thick. Since more of the field is in the copper, the conductivity has to *increase*. Thus, the decrease in the thickness of the aluminum cladding causes the conductivity of the article to increase.

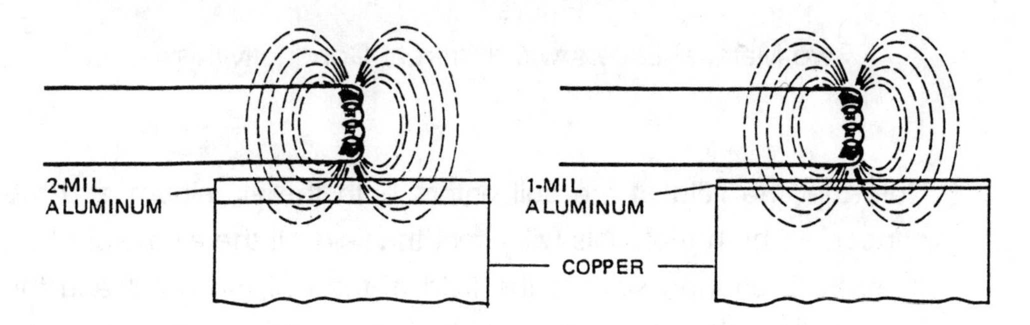

Figure 5-2
Magnetic Field Extension into a Two-Layered Sample

Now, reviewing the last few pages for a moment, you must realize that in some cases the addition of a conductive coating to an article will cause the conductivity of the article to increase, while in other cases addition of a conductive coating will cause the conductivity of the article to decrease.

- An *increase* occurs when the coating is *more conductive* than the base material.
- A *decrease* occurs when the cladding is *less conductive* than the base material.

The point is that the eddy current test technician has to keep his head out of the sand. He has to think about what he is doing, what he is working with, and the results that he expects to see so that he can recognize unexpected results.

We have now covered those five factors that affect measurement of the CONDUCTIVITY of materials. We have also explained the adverse effects caused by PERMEABILITY.

Now let's take a look at some DIMENSIONAL factors involved. We have already covered the dimensional factor of the spacing (lift-off) between the inspection coil and the material. Let's look at the ways in which the dimension and shape of a test specimen affect eddy current testing.

In Figure 5-3, article A, the material is so thick that the coil's magnetic field is completely contained in the material. In article B the same material is thin enough so that the magnetic field extends completely through the material; the point being that the part of the field not in the material does not cause eddy currents.

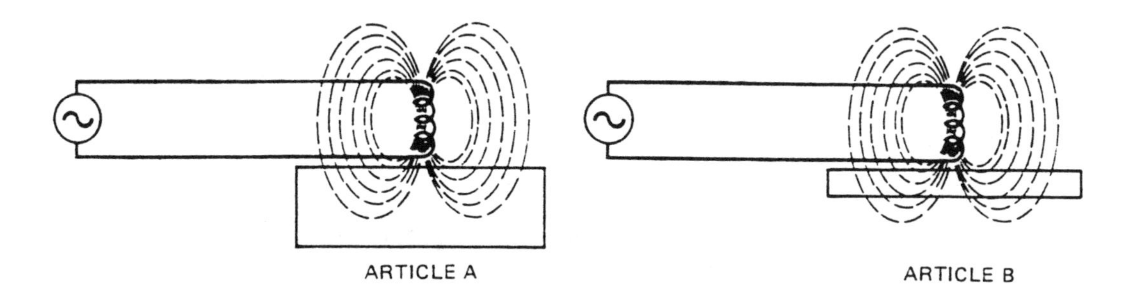

Figure 5-3
Conductivity Testing of Two Thicknesses of the Same Material

To sum up the discussion, we are saying that in eddy current testing for the thickness of thin materials, the equipment must be calibrated against *known thickness (standards)* of the material. Then, once calibrated for that particular test, the equipment, or settings on the equipment, cannot be changed without recalibrating. In short, all conditions of the test must be identical except for the thickness of the material.

Up to this point we have not emphasized the eddy current path in the material, but the paths that the eddy currents take do play a part in eddy current tests. As shown in Figure 5-4 the eddy currents *flow in circular paths* that are *parallel to the winding of the inspection coil*. They must form complete paths and, in thick materials, are *concentrated near the surface* next to the coil.

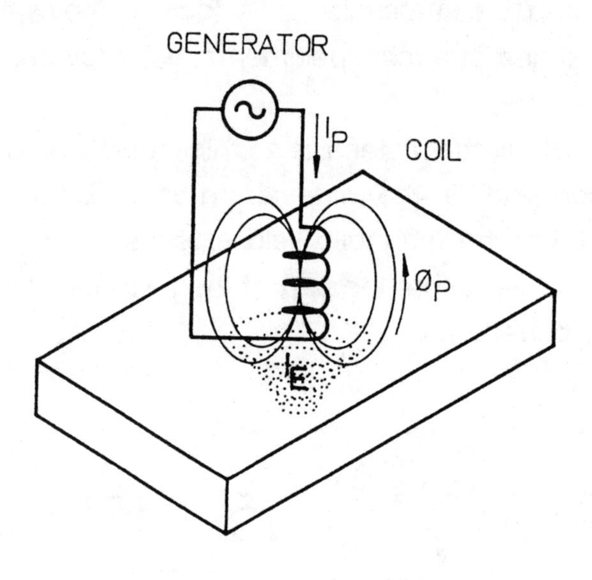

Figure 5-4
Eddy Current Flow in a Test Sample

If the material contains a discontinuity, as shown in Figure 5-5, the eddy currents are forced out of their normal circular paths. The paths become longer. Because the paths become longer, the currents are weakened, and the eddy current magnetic field is weakened. The crack will weaken the eddy currents and the secondary magnetic field also becomes weaker; therefore, the strength of the primary coil's magnetic field must increase.

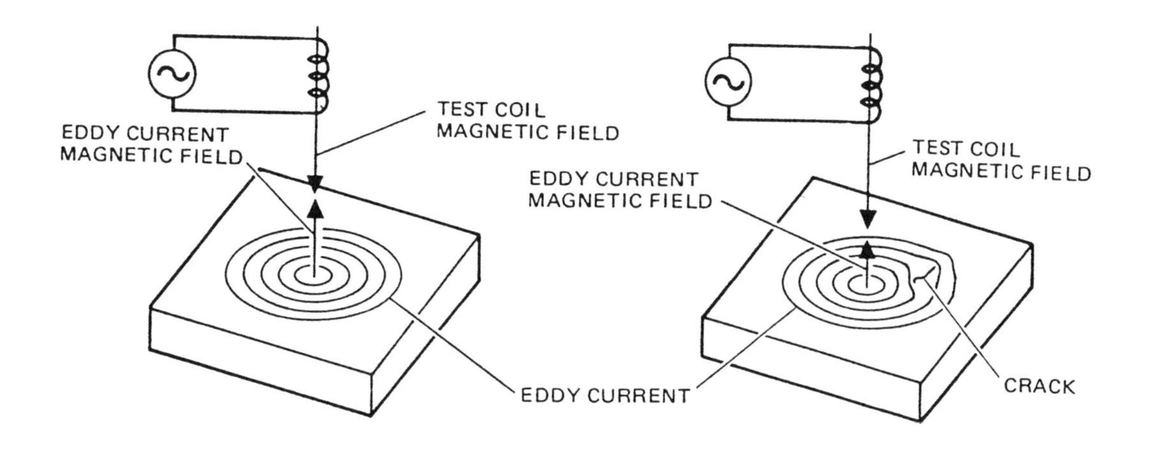

Figure 5-5
Changing Primary and Secondary Magnetic Forces
Due to Changing Material Characteristics

Figure 5-5 will help you understand how the presence of a crack causes the coil's field to increase. This diagram attempts to show the change in the magnetic fields due to the presence of the crack. The crack weakens the eddy currents by forcing them to take longer paths around the crack. Since the eddy currents are weakened, the magnetic field caused by the presence of the eddy currents weakens. Since the secondary magnetic field always opposes the primary magnetic field of the coil, weakening the secondary magnetic field allows the magnetic field of the coil to become stronger.

There is one more dimensional factor that you should be aware of. It is called the "edge effect." When the inspection coil is brought too close to the edge of an article, as shown in Figure 5-6, the circular paths of the eddy current become distorted. There is no material to support current flow. As the coil moves closer and closer to the edge, less and less of the current can flow.

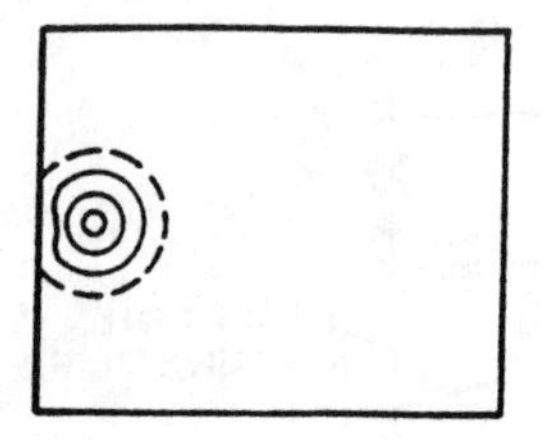

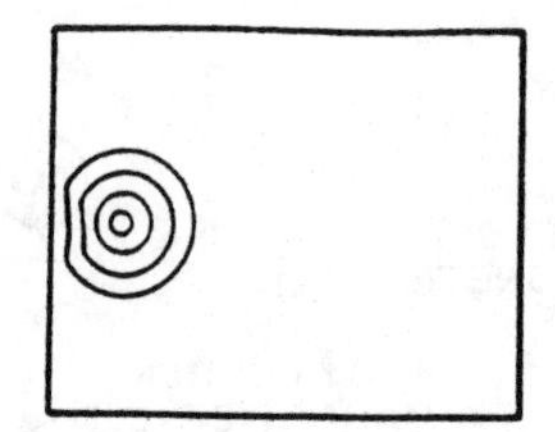

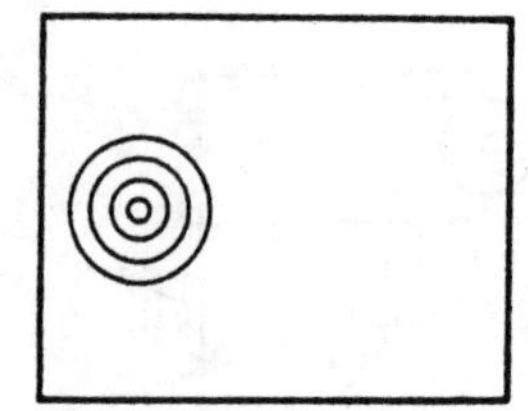

Figure 5-6
Distortion of Magnetic Field by Edge Effect

Since the edge of the article strongly interferes with the flow of eddy currents, edge effect is an important variable in eddy current testing. The limits as to how close to the edge a coil can be placed are determined by the size of the coil and how it is mounted. Locating the limit at which edge effect becomes a factor to be considered can be accomplished by approaching the edge with the inspection coil at several different places (to assure that the reading is not affected by some discontinuity) and observing the initial change in the instrument. This initial change marks the limit of coil measurement towards the edge.

The magnitude of the *edge effect will mask* any edge cracks that may exist, but it is not impossible to detect edge cracks because their effect on eddy currents is in addition to the edge effect. If the coil-to-edge distance can be held constant while the coil is moved *along the edge*, the effect due to the edge will remain at a constant value. A crack will cause a change above this edge-effect balance point. This can be accomplished by using a *special holder or fixture* so that the coil-to-edge distance is maintained at a constant value.

We showed in Chapter 3 that surface, encircling, and bobbin coils can all sense the material conditions within their geometries plus some area around themselves. The possibility of adding coil "shielding" to probe coils to limit this field spread and allow inspections closer to an edge were also discussed.

Keep in mind that it is the magnetic field extension around the coil that is of interest here. As shown in Figure 5-7, a rough estimate of an unshielded coil's magnetic field spread would tell us that a coil will "see" approximately twice its own diameter. This would mean that a probe coil 0.25 inch (6 mm) in diameter might be sensitive to material changes anywhere within a 0.5 inch (13 mm) "field of view."

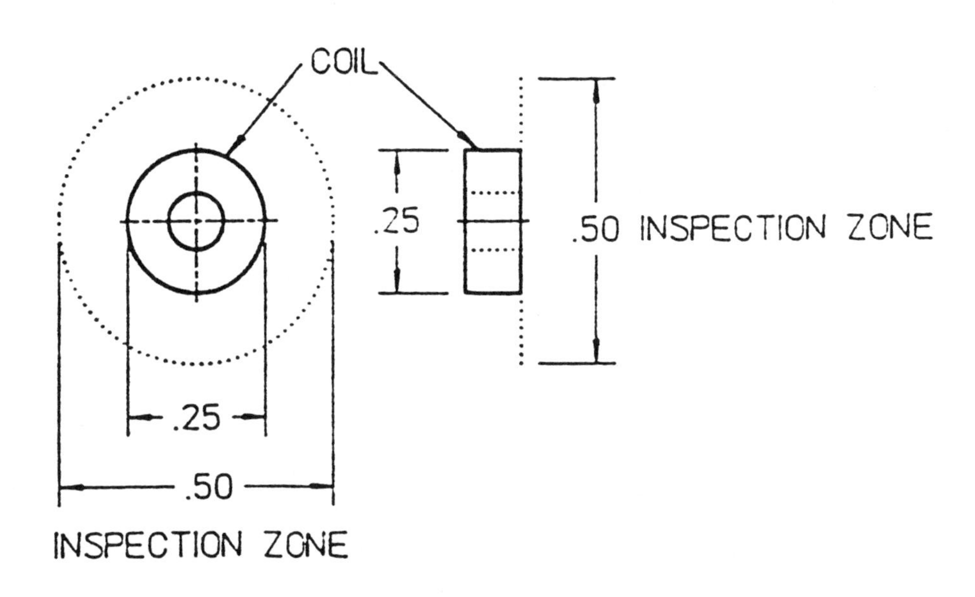

Figure 5-7
Magnetic Field Extension (ϕ_P) Around the Coil

For surface or probe coils this field spread can lead to the problem of edge effect. For encircling and bobbin coils, we call this same event "*end effect.*" All this means is that as the end of the tube or rod approaches the coil, it will be sensed prior to its actual arrival within the coil's windings. This sensitivity to a non-flaw condition could lead to a decreased ability to detect real flaws at or near the end of the inspection piece. Figure 5-8 illustrates how the end of a tube would cause a large amplitude signal due to the drastic decrease in conductivity; the total lack of conductive material where the tube ends.

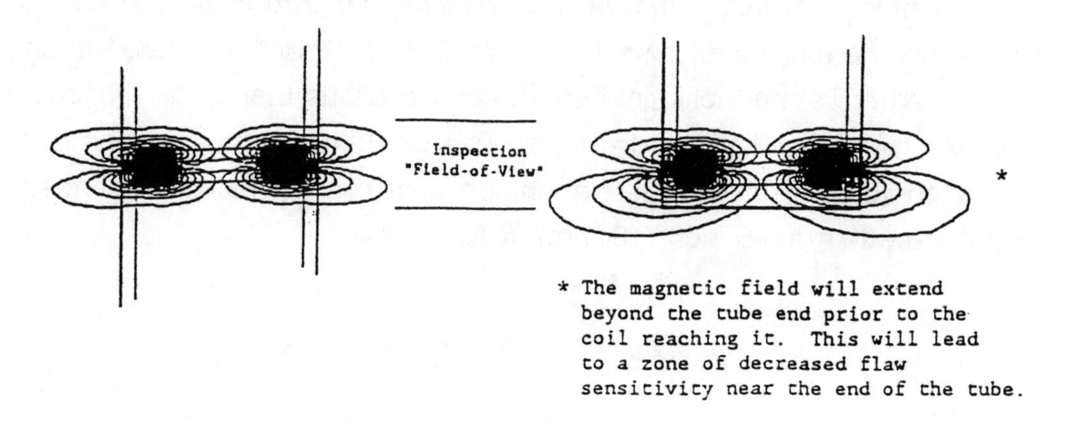

Figure 5-8
End Effect in Tube or Rod

A final dimensional characteristic that may affect the quality of our eddy current examination would be surface condition. Surface finish on castings and forgings can sometimes result in poor signal-to-noise ratios. Tubing may be produced by "pilgering" or extrusion manufacturing processes. If temperatures, feed rates, and pressures are not properly controlled, then minor ripples or thickness variations may be introduced into the tube wall or ID surface. These variations, although relatively minor, can lead to very poor test conditions.

Let's summarize the factors that affect the conductivity of materials. It is obvious that there are quite a few.

- The innate conductivity of the material can be used to identify the material, provided that the factors of *material hardness, temperature,* and *residual stresses* are either not present or can be suppressed; provided that the dimensional factors of *material thickness, lift-off, discontinuities,* and *edge effect* are either not present or can be reduced in the output signal response.

- The thickness of a given material ***can be gaged***, provided that the material is not too thick and that the other factors which affect conductivity are not present or can be eliminated.

- Cracks and other discontinuities can be detected, provided that the other factors which affect conductivity are not present or can be suppressed.

- Thickness of conductive or nonconductive coatings can be gaged (measured) by calibrating to lift-off responses, provided the other factors which affect conductivity are not present or can be reduced to a very low level.

- The eddy current test technician must be continuously aware of those factors which, if present, could affect the results of the particular test he is conducting. An important part of this job is to know which factors could be present and how to eliminate them or their effect from the test results. The technician must even be able to recognize when the factors are such that eddy current testing shouldn't be used.

We have also discussed the following relationships of eddy currents:

- The eddy currents always flow in circular paths that are parallel to the windings of the coil.

- The best response is obtained from discontinuities that offer maximum interruption to the flow of eddy currents.

- A discontinuity whose major axis cuts *across* the flow of eddy currents will cause the greatest disruption.

- The smaller diameter surface coils and the shorter encircling coils are more sensitive to discontinuities.

- The magnetic field of a surface coil is slightly larger than the coil itself, while the magnetic field of encircling and internal coils extends slightly beyond the ends of the coil.

- Eddy currents are most dense at the surface of a material and become progressively less dense with increasing distance below the surface.

- The standard depth of penetration is defined as the depth at which the current is approximately 37 percent of the current density that exists at the surface.

- The depth of penetration is affected by the conductivity and the permeability of the material; the higher the conductivity, the less the penetration and the higher the permeability, the less the penetration.

- Depth of penetration is affected by the frequency of the alternating current applied to the test coil; the higher the frequency, the less the depth of penetration.

- Approaching too close to the geometrical limits of a test sample can lead to edge-effect or end-effect. These false indications could distort our eddy current test signals.

CHAPTER 6

TEST SYSTEM CALIBRATION

Up to this point, our discussions have covered the basic eddy current systems, type of coils, and how certain physical properties of the material will interact with the coil and its induced eddy current field. Before we can attempt to do any real testing, we have to look at one additional parameter.

In order to understand the eddy current output response signatures (or signals) due to material conditions, we have to be able to compare them to something "known." When we see a particular signal on the screen, we have to determine if it is a flaw or a response that is acceptable in the test piece.

Now let's try something. Let's balance the system with the coil on a piece of copper. Now we move the coil along the piece of copper and suddenly we get a momentary screen trace or needle deflection. The response seen was probably caused by a discontinuity. But to be sure that it was caused by a discontinuity, we must be able to eliminate the other factors that might cause a similar change. "Lift-off" would have given the same type response. If it is "lift-off," a second pass over the same area, if made with care, will eliminate the deflection.

Discontinuities have the characteristic of causing a sudden deflection with an immediate return to zero except when the coil is passing along a discontinuity such as a seam. In this case, moving the coil crosswise to the line of the seam will provide more rapid response and thereby identify the deflection as being caused by a seam.

Now we are about to test a thin sheet of aluminum to see if the thickness is within tolerance over the entire area. Our first step would have to be deciding what our acceptance limits are going to be. The technician then would have to obtain standards made from that type of aluminum and milled to the exact upper and lower limits of the allowable thickness tolerances. You have probably guessed that in this case it is not absolutely necessary to balance the bridge so that we get a zero reading for one of the limits.

So this brings up the question, Is it absolutely necessary to balance the bridge so that one of the factors of interest will give a zero reading on the system? The answer is, No, it is not necessary. In fact, the point at which the system is balanced (called the test point) is used to differentiate the causes of readings. This subject will be covered in more depth in Chapter 7.

Let's think about our basic circuit for a minute. In Figure 6-1, we show two identical test coils used as impedances in the bridge.

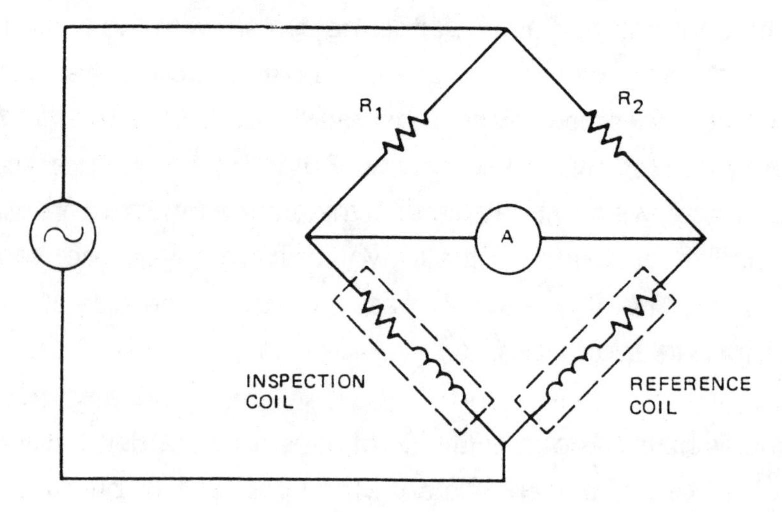

Figure 6-1

Two Identical Coils in a Bridge Circuit

The bridge will be in balance if the coils are identical in every way, since being identical would include having identical impedance. There is an obvious advantage in this arrangement. The advantage is that the bridge will normally be in balance. The two coils are simultaneously used as inspection coils and an imbalance in the bridge indicates a difference in the materials under the two coils.

For example, if the coil were placed on a reference standard for a particular test and the inspection coil were placed on one of the items being inspected, the meter would indicate any differences between the reference standard and the item under test.

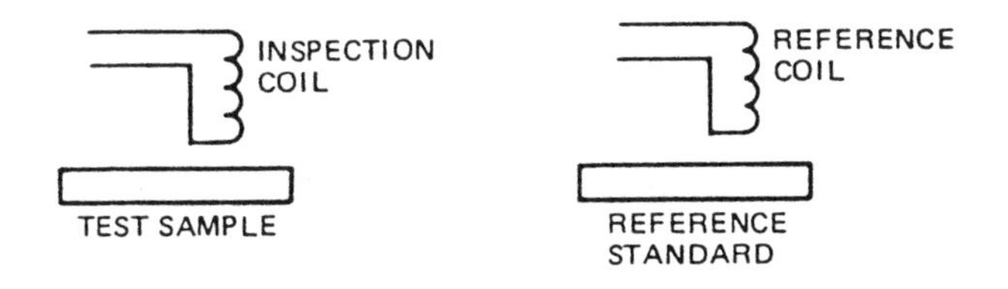

Figure 6-2
An External Reference Coil Arrangement

Another variation in the use of an inspection coil and a reference coil is shown in Figure 6-3. The two coils are mounted in a fixture as shown, then passed over the material. The impedance of each of the two coils depends on the condition of the material under that coil.

The two coils are held a specific distance apart by the fixture. As the fixture (and the coils) is moved over the surface of the material, as indicated on the next page in view A, coil A will pass over the discontinuity while coil B is still over sound material. An imbalance results which is seen as a response on our display. A little later, coil B is over the discontinuity while coil A is over sound material. Again, a response shows an imbalance, but in the opposite direction. In this case, the two coils are comparing flawed material with *sound* material.

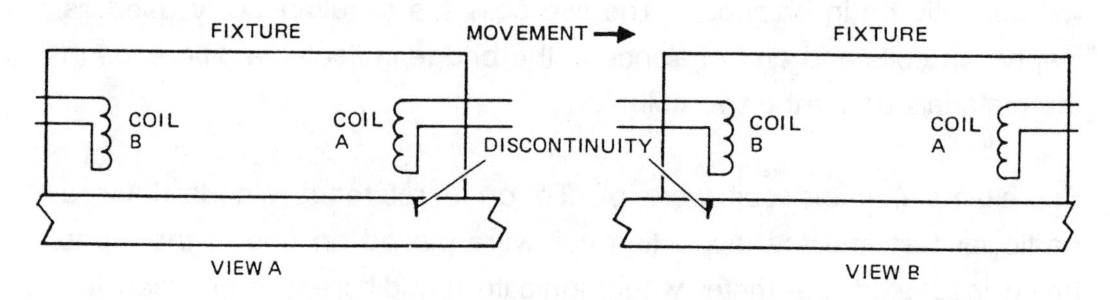

Figure 6-3
Two Coils Mounted in the Same Fixture

Standards

As in other types of nondestructive testing, the most useful test data is obtained by comparing the data from an item under test with data obtained from a reference standard. Standards furnish an exact value that has been established by authority, custom, or agreement as the norm by which other like articles may be judged. Standards also help in the design of procedures developed to measure those quantities that are represented by the standard. Standards often define the limits of acceptability of an item and serve to ascertain that the equipment being used is capable of measuring that quantity to the required degree of accuracy. A standard is also used to make sure that the equipment provides consistent sensitivity each time the equipment is used.

If fabrication of eddy current standards is required, it is normally left up to a company or group that specializes in that service. The machining techniques, precision measurement, and documentation (traceable to NIST, etc.) involved normally are more than the typical machine shop can provide.

Conductivity Standards

Metal blocks representing specific values of conductivity are typically supplied with conductivity testers. The blocks required for calibrating each system should be procured based on the applications and procedures governing each test. The standards used must be stamped with their conductivity values in percent IACS and should fall within the range and precision requirements for each test.

- If the test requires accuracy across a *broad range* of conductivities (**8 percent to 110 percent**), then the required number of calibration blocks used should encompass the full range (i.e., 16 percent, 42 percent, 60 percent, and 100 percent IACS).

- If the test requires the same precision or accuracy across a *medium range* of conductivities (**25 percent to 65 percent**), then the procedures used may require the same number of calibration blocks as before but they would now have to fall within a narrower band (i.e., 16 percent, 29 percent, 42 percent, and 60 percent IACS).

Keep in mind also that these system calibration blocks may themselves have to be "calibrated." Some materials will undergo a natural aging process that actually makes them harder. You will remember that grain structure and hardness were discussed as variables that can affect the conductivity of a material.

So how does one calibrate a piece of metal? Calibration, in this sense, would mean that the conductivity blocks would have to be measured with an independent "master" system and compared to a "master" set of calibration blocks. The blocks themselves would then be "certified" as acceptable to be used as system calibration blocks. The system calibration blocks supplied with the eddy current tester might be permanently attached to the unit. In that case, the blocks would be measured and their conductivity verified as acceptable when the instrument goes through its normal calibration cycle.

Another option is calibration block holders as shown in Figure 6-4. These also must undergo a certification procedure but could be used with more than one eddy current tester.

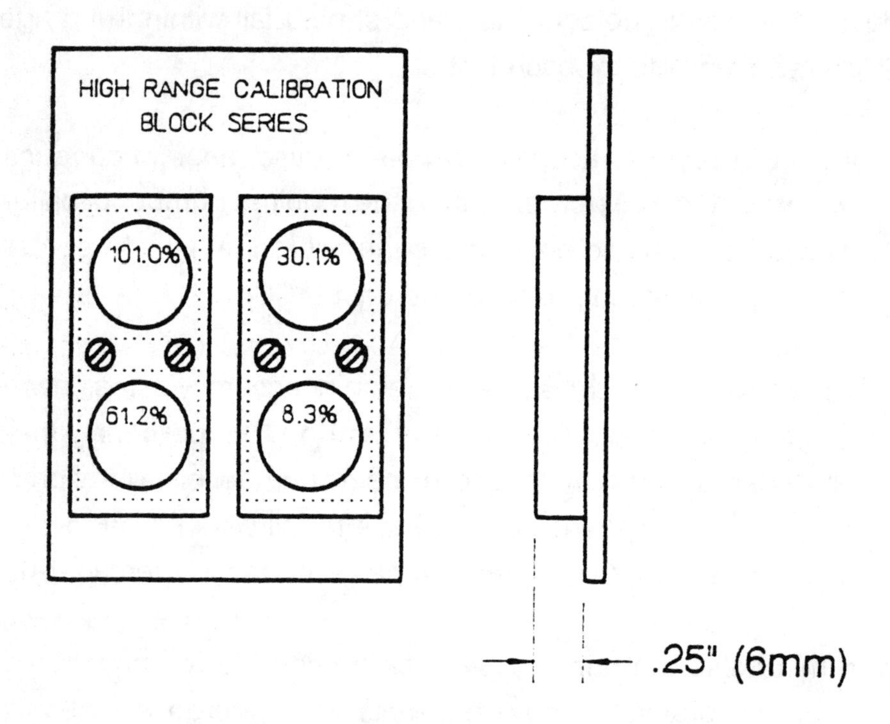

Figure 6-4
Conductivity Block Calibration Set

The conductivity tester is always adjusted to read the value *stamped on the block* when calibrating the instrument. Since the "high" block has 101.0 stamped on it in Figure 6-4, you know that the conductivity of that block is exactly 101 percent IACS. If the system is adjusted to read 101 percent when the test coil is placed on that block, then the "high" end of the system is calibrated.

The next step is to calibrate the system so that the low end of the readout mechanism will show the value stamped on the lowest block required, in this case 8.3 percent IACS. After these two steps have been accomplished, the system has been fully calibrated and is ready for use in the high-range test situation.

Discontinuity Standards

Ideally, a discontinuity standard should duplicate the test situation as closely as possible. Duplication of the test situation includes material type and geometry as well as duplication of the type of discontinuity sought. The reason that the geometry of the standard should be the same as the geometry of the test articles is fairly obvious for pieces that are not flat. Geometry is also very important for thin pieces, since thickness in those ranges has such a large effect on the results. Obtaining samples of test articles for use as reference standards is not always easy.

Discontinuity standards fall under two types—natural and artificial—depending on their source.

- Natural Discontinuity Standards

  Natural discontinuity standards consist of duplicates of the test piece configuration that contain discontinuities of a known size and shape that have occurred from natural causes. Natural discontinuity standards can be developed or accumulated. By submitting a test sample to cyclic stresses, a "natural" fatigue crack can be produced in the sample. Since we have taken action to deliberately introduce a discontinuity into the test sample, the discontinuity is also a "developed" discontinuity. It still is defined as a "natural" discontinuity, since cyclic stresses could be naturally applied when the part is in service.

An accumulated discontinuity is one which might occur during the manufacturing processes or inservice stresses applied to the part. Articles that contain this type of discontinuity may be accumulated over a period of time during routine testing of articles. Samples containing natural discontinuities, either developed or accumulated, may be machined to produce a surface crack or hole of a known depth, as shown in Figure 6-5.

At least one of the cracks in the reference standard should be at the limit of acceptability. This means that we will reject anything that has a crack that is bigger than this one.

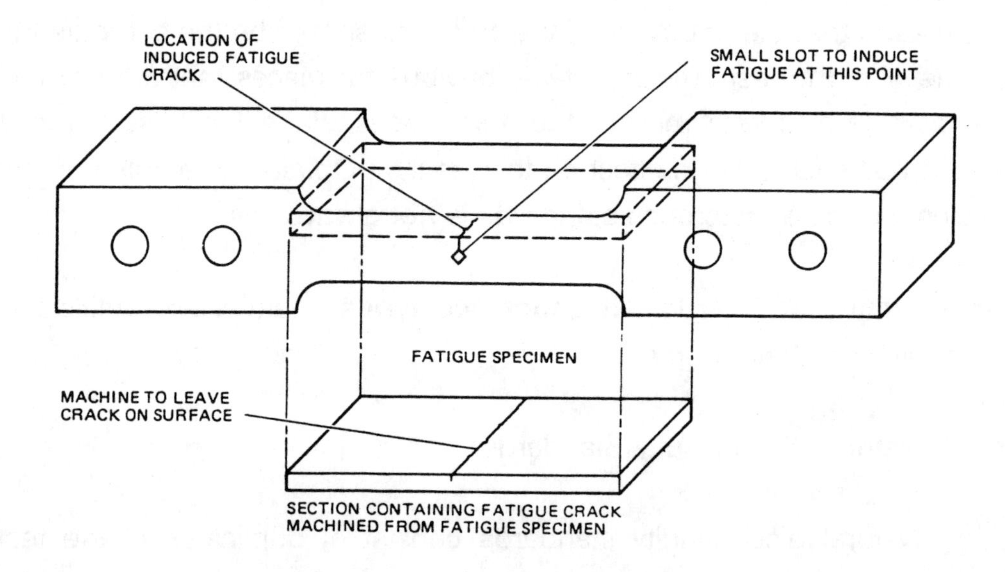

Figure 6-5
A Process for Developing a Crack Standard

- Artificial Discontinuity Standards

Artificial discontinuity standards are standards that are prepared by machining artificial discontinuities into an article that has no natural discontinuities. Several samples may have to be run through the inspection system to find one that does not produce any appreciable indications of natural discontinuities.

Once such a sample is located, standard reference discontinuities that are pertinent to the required specification are then fabricated into the sample. Types of standard reference discontinuities used to simulate natural discontinuities are longitudinal notches, circumferential notches, drilled holes, diameter steps, and geometry variations. The reference standard should have artificial discontinuities that represent the limit of acceptability.

There is a test situation, however, where we need a standard that represents perfection. When external comparison techniques are being used, the standard used with the reference coil should represent perfection for that article. It must be free of discontinuities. However, even then the limits of acceptability must be established by using a reference standard containing the required discontinuities within the field of the test coil to obtain a reading that represents the limit of acceptability.

Tubing Standards

The procedure or specifications controlling surface inspection standards may only require one discontinuity, perhaps a crack or notch of a known width and depth, to represent the "limit of acceptability." The standards required in other types of inspections to meet the acceptance criteria may require that we accurately report the actual size of the discontinuity. In this type of inspection, we will have a standard with multiple discontinuity orientations and sizes to provide the possibility of a much more comprehensive data analysis.

An extended range of artificial discontinuities permits a more detailed signal interpretation process. The proper use of these specialized standards can allow a skilled eddy current technician to begin to make distinctions between signals arising from *discontinuities* (acceptable damage) and *flaws* (discontinuities that exceed an engineering limit).

Eddy current testing is widely used in the inspection of tubing. The first EC testing is normally done in the mill during the manufacturing stages. A second exam might be performed after the individual tubes have been assembled into the final product or when the product is delivered to the end user. Furthermore, eddy current testing will be performed at given time intervals after the finished product is put into use. This type of testing provides a data base of information about the original and present state of each tube in the heat exchanger and whether or not it is still safe to use.

The type of material, the material geometry (bent, straight, expanded, finned, etc.), and the heat exchanger operating characteristics normally will tell us what kind of damage may be present. Our choice of probes and standards are controlled by these same features. We need to keep in mind that any standard is going to have limitations as to how and when it can be used. Each standard is designed to provide information about a certain type of damage when seen with a certain type of probe.

A typical tubing calibration standard has an array of artificial discontinuities, as shown in Figure 6-6 in this sample of 90/10 copper/nickel (Cu/Ni) alloy.

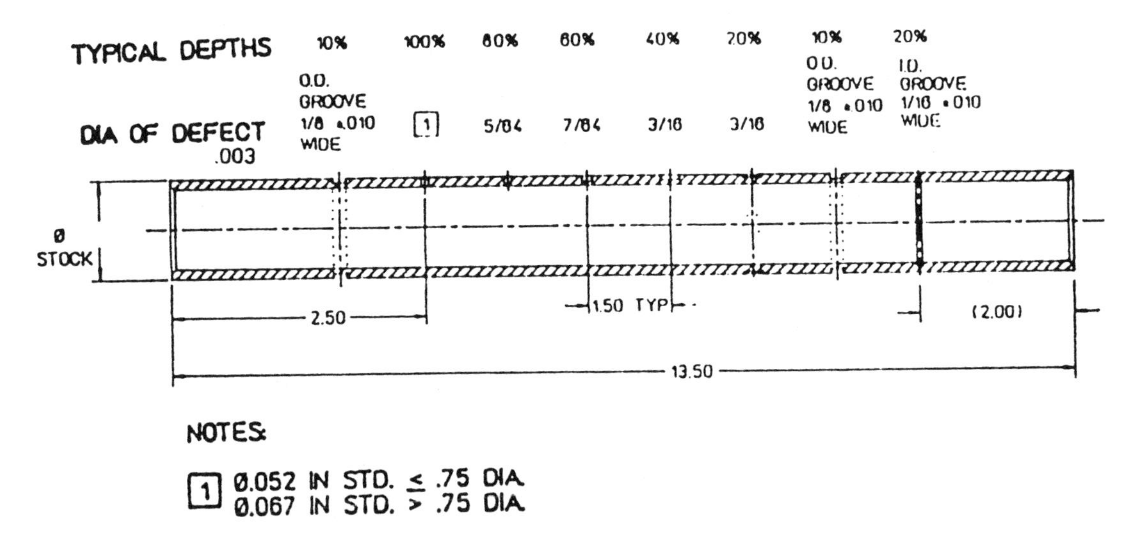

Figure 6-6
A Selection of Discontinuities in a Calibration Standard Tube

To perform a good inspection of any material, we have to have a standard that exactly models the material that we want to test. In choosing our standards, there are certain things we must remember.

- Any variation in either the material or the test geometry will create a change in the strength and distribution of the eddy current field (Chapter 2).

- A change in either of these variables is going to create a change that affects the way the two magnetic fields (primary and secondary) relate to each other and therefore alter the coil response (Chapter 3).

- A change in operating characteristics of our coil is going to be sensed by the test system as a voltage shift across the bridge circuit (Chapter 4).

- If we do not control the "known" variables in the test situation, then we will have very little chance of understanding the complex patterns of possible responses in the test results (Chapter 5).

Let's look at some of the possible variations in our tubing standards. We mentioned the different types of possible artificial discontinuities that might be used in creating standards earlier in this chapter. Let's expand on this.

- *Notches*, either circumferential (around a tube) or axial (down the length of a tube), can be created with great precision by a process known as electrode discharge machining (EDM). All dimensions of the notch, as well as critical placement, can be very accurately controlled. With modern equipment and a skilled operator, these types of artificial discontinuities can actually be placed on the inside surface of tubing specimens to model ID discontinuities.

- *Drilling*, either using standard drills or special milling tools, can create artificial discontinuities that are 100 percent through-wall (TW) or are at very precise increments of the wall (20 percent, 40 percent, 60 percent, etc.). The milling tools provide the capability of making flat-bottomed holes (FBH) with known depth and volume. This allows for output signal calibration techniques which are very repeatable (more on this in Chapter 7).

- *Diameter steps* can be created on a lathe to simulate volumetric material loss as might be found in a fretting or erosion/corrosion type of situation. With care, these steps can be created on either the outer (OD) or inner (ID) surfaces of a tubing standard.

- *Geometry variations* can be created by a controlled expansion (bulging) or compression (denting) of the tube wall. These can model either an intentional design condition or an inservice generated condition.

In order to accurately model a given field situation, it may be necessary to create combinations of these events in a reference standard. Only by knowing the total combined signal pattern can we start to control the variables that will allow us to perform *signal suppression techniques*.

Standards allow us to see how "known" signal sources, whether flaw-related or simply a typical tube heat exchanger characteristic, are going to respond during the examination. If we can suppress (reduce, eliminate, or mix out) the signals that we know are not discontinuity-related, then the only output signal information remaining to analyze must be flaw-related.

The tubing test standard shown in Figure 6-7 on the next page is referred to as an American Society of Mechanical Engineers (ASME) standard. The main type of artificial discontinuities used are "flat-bottomed holes" (FBHs). They simulate localized damage because they are not very long and do not extend around the tube circumference. These discontinuities provide signals similar in nature to various cracks and/or pits found in service-damaged tubing. This type of damage typically has a very quick eddy current test output response, hence the name "fast rise time" or high frequency responses. The damage might be located either in free span tubing or where a tube passes through a baffle or support structure.

In Figure 6-7, notice the ring of metal around the tube that simulates one of these support interfaces. This simulated tube support ring (TSP) is always situated in a nonflawed segment of tubing, never at or near an artificial flaw.

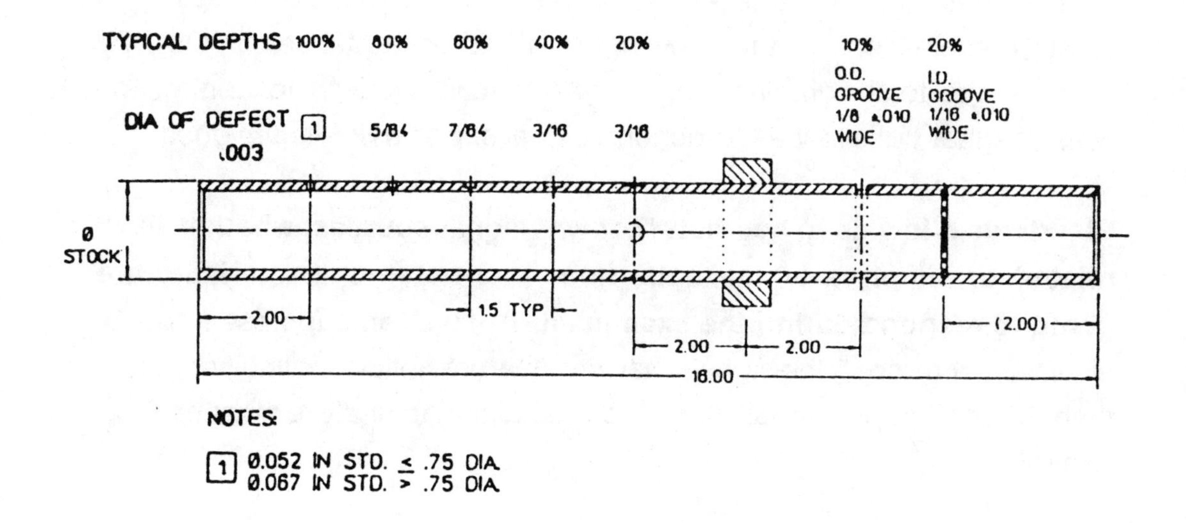

Figure 6-7
ASME FBH Standard (w/TSP)

Figure 6-8 shows one possible model for a wear or fretting condition of the tubing. This discontinuity type is one of those that is normally addressed as "volumetric" wall loss. It can normally be attributed to vibration of the tube where it passes through a support structure. Note that the artificial discontinuities are much broader in this standard. This would give them a much longer output response on the screen than the small FBHs we saw in Figure 6-7. These types of signals could be called "slow rise time" or low frequency responses. Also note that there is one nominal TSP provided and that the other damage sites are all under the other TSPs provided.

Another version of a volumetric damage mode might be called a wall thinning standard, as shown in Figure 6-9. This damage is not going to occur under the TSPs, so the standard does not provide them.

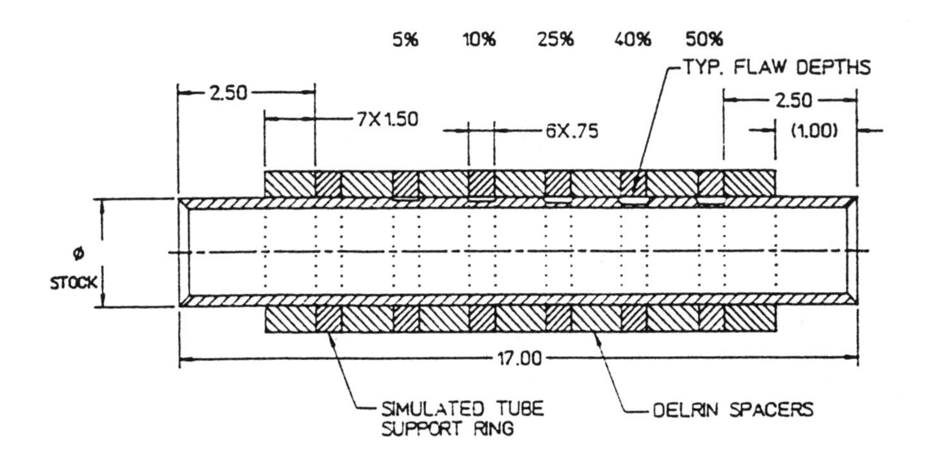

Figure 6-8
Wear Scar Standard (w/TSP's)

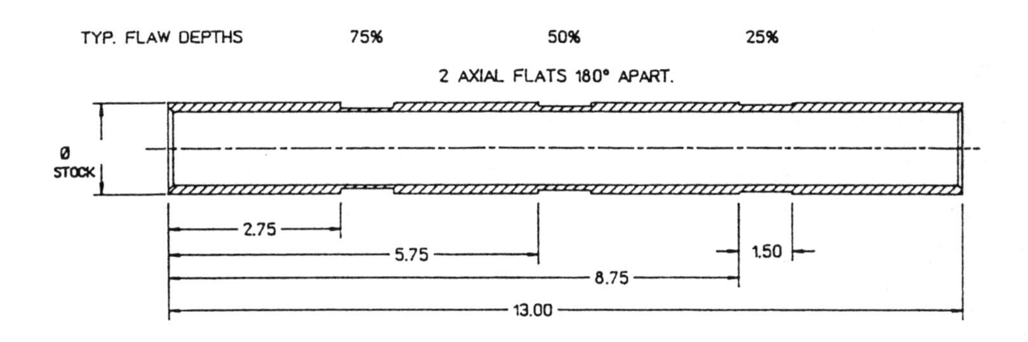

Figure 6-9
Wall Thinning Standard

We are almost ready to go out and start testing! The greatest drawback to any test is overconfidence by the user about its potential performance. We can see from our general discussions that the decisions that we make in the analysis process *are only as good as the standard(s) that we use to set up the examination.*

We cannot realistically say that there are **NO** discontinuities or flaws in the material we are inspecting. The **ONLY** statement we can make is one that summarizes what we really know.

> "Based on the equipment, frequency(ies), probe(s), and standard(s) used, there were no output signals **detected** that *exceeded the acceptance criteria.*"

It is critical that we choose a standard that will ensure that we have good flaw "detection" capability. Some of the elements that need to be considered to assure proper selection of tubing standards might include those listed below.

- Some knowledge of the basic material and its engineering specifications
- The design and operational characteristics of this type of unit or heat exchanger
- The "typical" damage mode (type of flaw, location, and orientation) for this material under the given operating conditions
- Any past history on this unit or similar units

Aircraft Structures

Another sector that has found frequent applications for eddy current testing is the aircraft industry. This is true for the manufacturing organizations but even more so where environment and stress have led to "aging aircraft" issues. As we stated earlier, the standards that we use must accurately model the actual test specimen. Ideally, we would like an actual part with "known" or controlled discontinuities; but, since it is difficult to quantify these flaws, the next best thing is a sample with similar part geometry and conductivity characteristic with artificial flaws. This will allow us to verify that our probe and instrument combination are capable of detecting the discontinuities determined to be critical to each application.

Figure 6-10 shows a multipurpose standard that would allow us to calibrate our system for:

- detection of near-surface cracks on a flat surface,
- detection of near-surface cracks in a curved radius,
- detection of sub-face flaws on a flat surface,
- detection of flaws inside a bolt hole, with and without the fastener,
- verification of material conductivity.

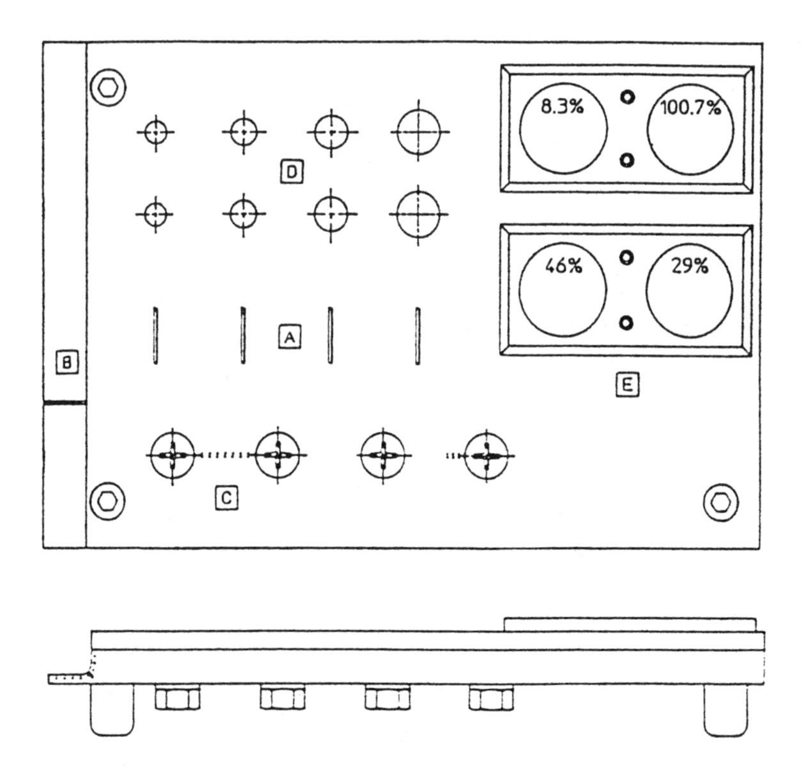

Figure 6-10
Multipurpose Aircraft Standard

Lift-off Standards

Since lift-off amounts to having a nonconductive space between the test coil and the article, lift-off standards are easy to construct. The application of a known thickness of any nonconductive material to a sample of the material under test will constitute a lift-off standard. Paper, mylar, and cellophane are examples of nonconductive materials often used. If we are measuring the thickness of a nonconductive coating over a conductive article, we need to construct lift-off standards that represent both the maximum and the minimum acceptable thickness.

As shown in Figure 6-11, layers of paper, mylar, or other plastics may be built up to the required thickness for the standard. As we have already stated, reference or "calibration" standards are used to correlate the readings from the test piece to conditions that we know exist in the reference standard.

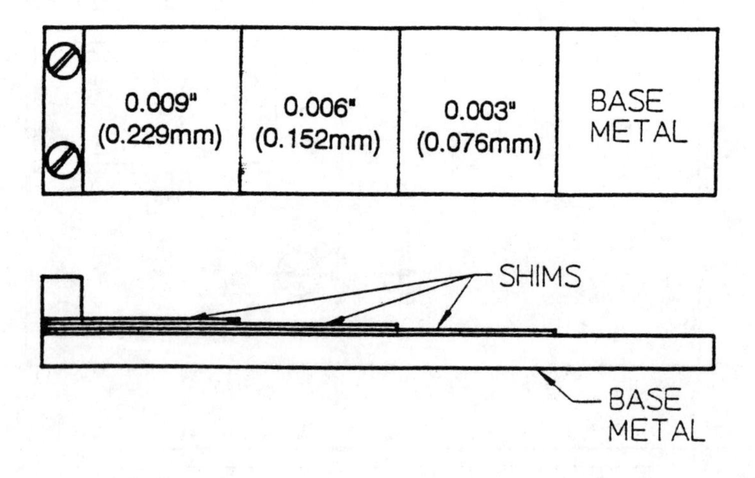

Figure 6-11
Nonconductive Shim Stock Used as a Lift-off Standard

Equipment Calibration Verification

Standards are also used in another way. After the equipment has been calibrated to the standard, the electrical settings of the tester should not be changed. If this happens, the test results will not be accurate. For this reason, it is wise to recheck the equipment against the reference standard whenever an unexpected result is obtained. This will help assure that the cause of the unexpected reading is not due to a fault in the equipment.

During long production runs, it is wise to periodically recheck the instrument against the reference standards to make sure that the electrical characteristics of the test set have not "drifted," thus causing erroneous test results. If the equipment calibration verification is found to have drifted, and the system no longer meets the minimum detection criteria, then it may be necessary to retest all material since the last system check.

Summary

- **Balance Point**—The use of bridge circuits allows us to "null" or cancel out a certain offset voltage or response. Our choice of balance points (either in air, on a reference sample, or on a test sample) will depend on what we want to establish as our "zero" point and what we want to know about the material we are testing.

- **Calibration**—Once we have established this balance point, we must then understand how other changes will look on our test output device. In order to do this, we must have different types of standards.

- **Standards**—Conductivity, discontinuity (both natural and artificial), and lift-off standards were discussed. Some standards might be defined as "industry specific," such as aircraft or tubing standards.

The range and type of information that is required as an end result of the inspection will dictate the types of standards required.

CHAPTER 7

IMPEDANCE-PLANE CONCEPTS

In the preceding chapters we have explained the factors that affect the eddy currents induced by the inspection coil and their effects on the test results. In this chapter we will explain how to determine which variable is causing the change in the output response.

Eddy current testing is fairly simple to perform. The most difficult task for the operator is to be able to judge (with any degree of certainty) which variable caused the change in test coil impedance. For example, we have learned that a change in the hardness of a material will affect its conductivity. We also know that the presence of a discontinuity will also affect the conductivity of the material.

We can predict that a gradual change in conductivity is due to a change in hardness and not a change due to the presence of a discontinuity for two reasons. First, the change is gradual (unlike the change due to the presence of a localized discontinuity); and second, the material under test, in this case, has a history of applied heat friction.

In this same manner, simple reasoning can differentiate changes that might occur because of temperature differences or because of the presence of internally-stressed areas. The real problem is to differentiate between changes in impedance that are due to conductivity factors and changes that are due to lift-off or fill-factor.

The paragraphs that follow present detailed methods and techniques that are available through the use of more sophisticated test instruments. As you know, the impedance of a coil may be represented by a vector whose length represents the impedance value and whose direction represents the phase angle (the angle by which the current lags behind the voltage).

Looking at Figure 7-1, we could make the following statements:

- The current through the inspection coil lags behind the voltage applied to the coil by 37°.
- The impedance of the test circuit is *5 ohms*.

Now that we understand the meaning of the impedance vector, we must clearly understand that the impedance vector in Figure 7-1 is made up of **two components**: the resistive (R) component and the inductive reactance (X_L) component as shown in Figure 7-2.

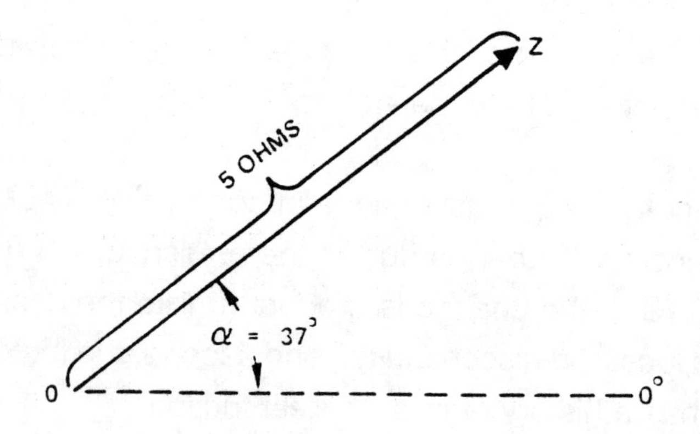

Figure 7-1
A Vector Diagram Showing Phase and Amplitude

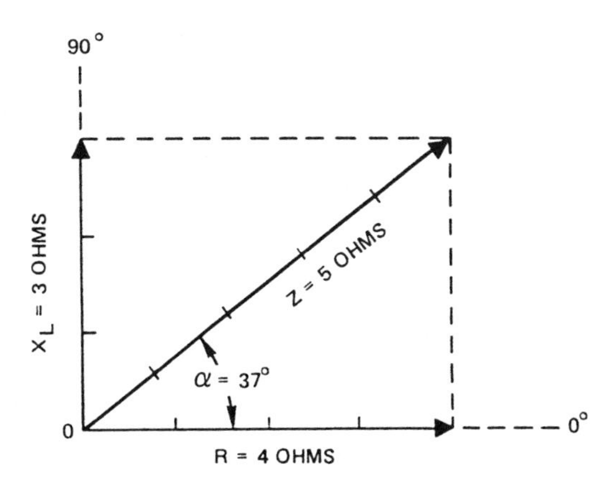

Figure 7-2
A Vector Diagram Showing Impedance Vector Values

To determine the value of these two components we draw a vertical dashed line from the impedance value to the horizontal (X) axis and a horizontal dashed line to the vertical (Y) axis as shown. In this manner, we have determined that the resistance value is 4 ohms and the inductive reactance value is 3 ohms. By using our impedance formula or by plotting the impedance on a graph, we can determine the impedance once we know R and X_L.

What has all this got to do with eddy current testing? Just this—as we have said countless times, the conditions that exist in the material under test affect the *impedance* of the test coil. So knowing the impedance gives us a tool for understanding the relationship between the test coil and material factors that affect the conductivity. We can use the phase angles and amplitudes of these vectors to create "impedance-plane" diagrams.

Let's take several kinds of material, making sure that they are thick enough so that the thickness (thinness) does not affect the readings, and measure the impedance and phase angle produced in the test coil as it is passed over each piece. The result will look something like Figure 7-3.

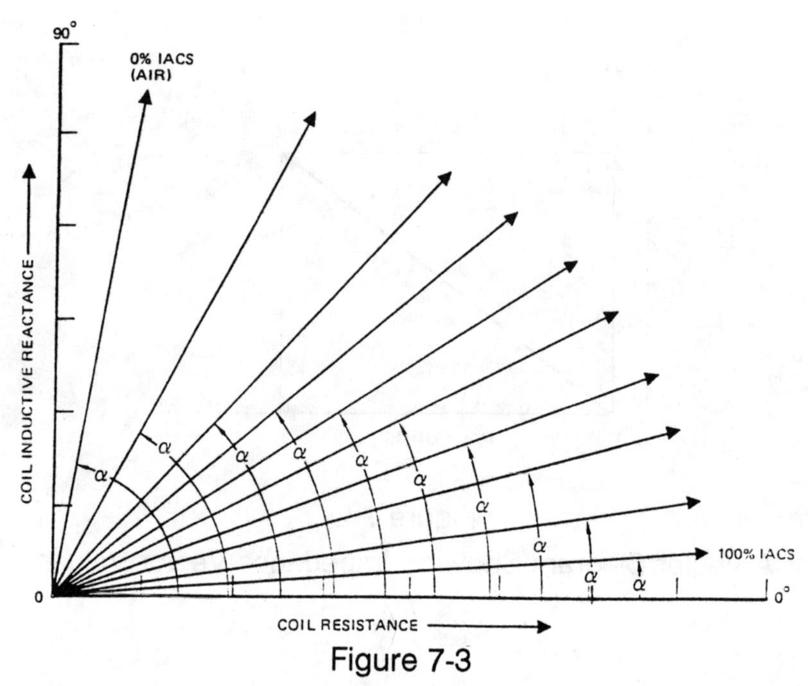

Figure 7-3
Impedance Variations Based on Material Conductivity

If we have been careful in selecting our material and in taking the measurements, the only thing that has varied from measurement to measurement is the conductivity of the different materials.

In Figure 7-3 we have plotted the impedance of all kinds of material having conductivities ranging from air at 0 percent IACS to copper at 100 percent IACS. Now let's draw a curve that connects all of the impedance values that we have obtained. The result will look like Figure 7-4.

The conductivity of any nonferritic metal will cause the impedance of the test coil to fall somewhere on this curve. The curve is a line drawn through all of the measured impedance values that will result from changes in conductivity for this particular test setup.

If all other factors are held constant, a change in conductivity will result in an impedance value that will fall somewhere on the curve in Figure 7-4.

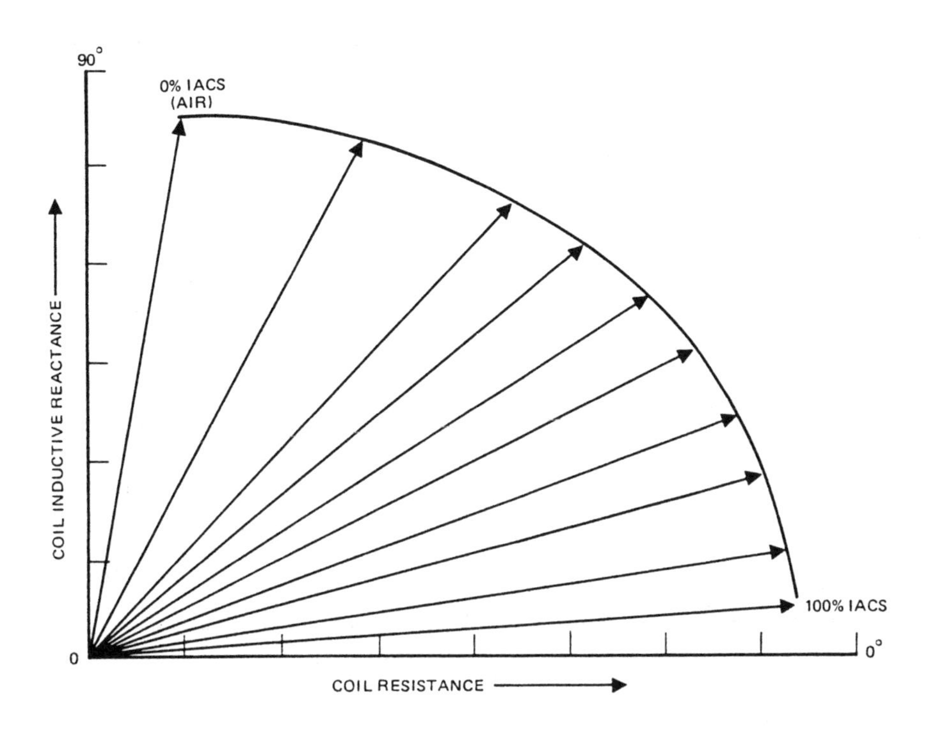

Figure 7-4
Impedance Variations Defining a Conductivity Curve

In Figure 7-5 we have identified the material as bronze. If we changed the alloy chemical content or temper and it caused a decrease in the conductivity of the bronze, the impedance point would move up and to the left on the curve.

Remember that this conductivity curve is the result of a particular test setup. Any change in the test setup will result in different values of impedance being obtained. The different values of impedance will, in turn, result in a slightly different curve. However, the curve we have shown is representative of all conductivity curves.

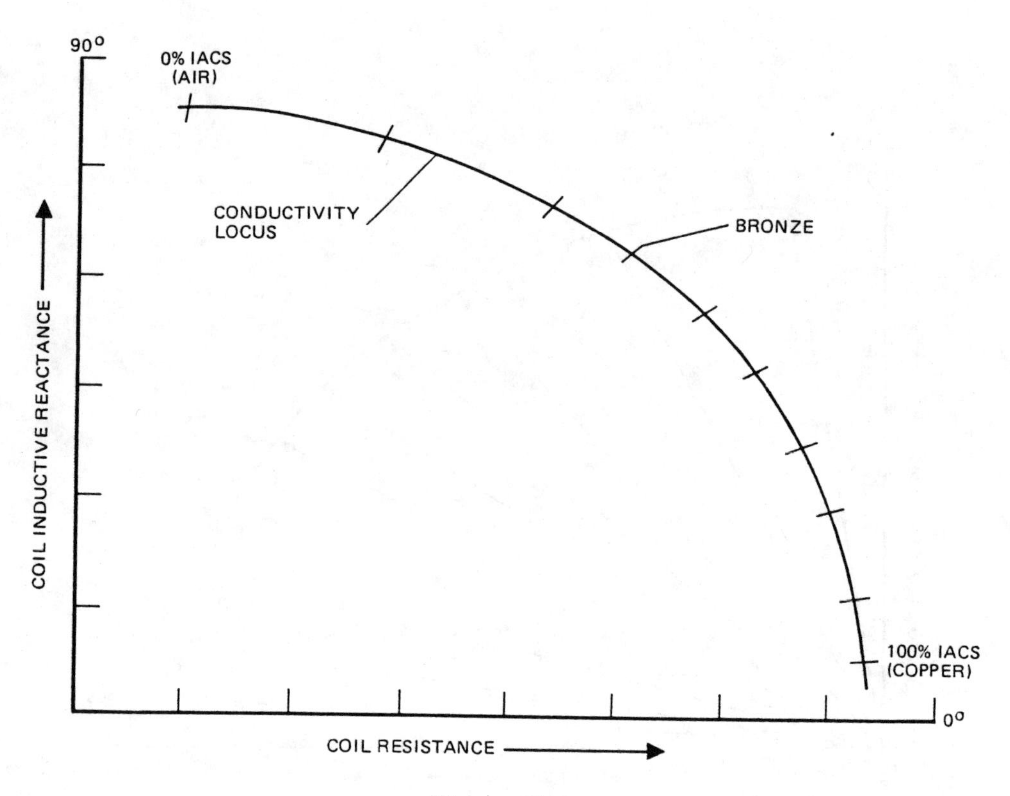

Figure 7-5

A Conductivity Curve Showing Varying Values of Percent IACS

Now let's see how lift-off looks on the impedance-plane diagram. When the test coil is held in contact with the 100 percent IACS material, the impedance value will fall on the conductivity curve. As the coil is gradually lifted off the material by means of paper shims placed between the coil and the material, the impedance moves in the direction of the dashed line as shown in Figure 7-6.

Note that as the coil is lifted more and more that the lift-off curve finally meets the conductivity curve at the 0 percent IACS point. The impedance of the "unloaded" coil is measured by the vector from point A to the 0 percent IACS point. The test coil always has some impedance, even when it is held in air.

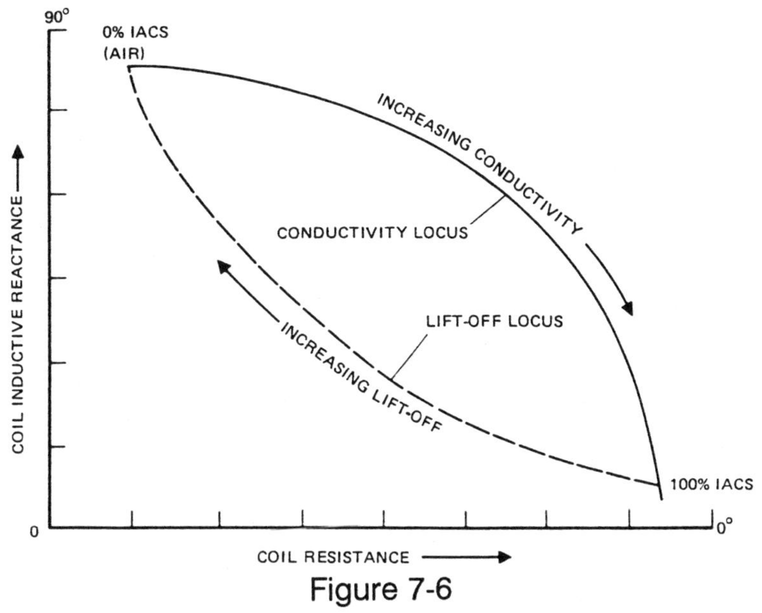

Figure 7-6

A Conductivity Curve Also Showing Lift-Off

Keep in mind that the lift-off curve we have just described is for one material only—the 100 percent IACS material. What is the significance of the lift-off curve on the impedance-plane diagram? Let's look at just part of the diagram in Figure 7-7.

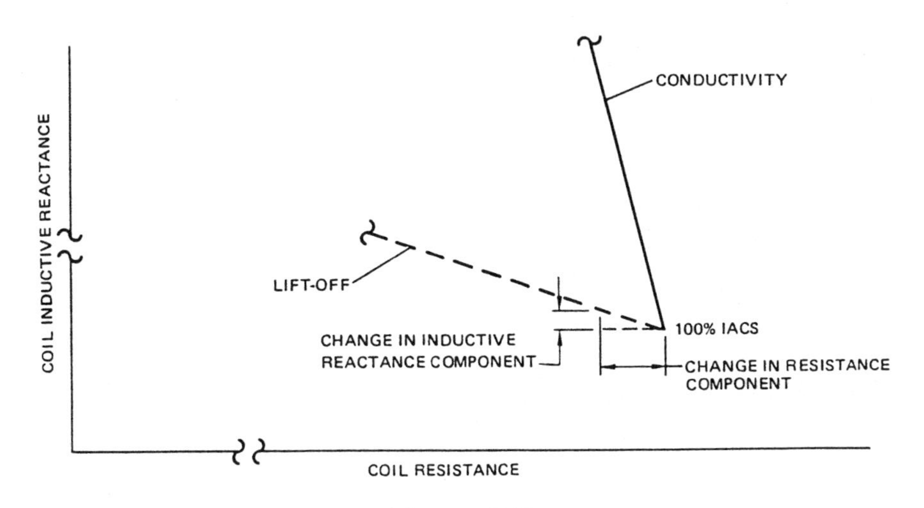

Figure 7-7

A Small Segment of the Conductivity Curve

Notice that the effect of changes in lift-off are in a different direction than changes in conductivity on the impedance-plane diagram. This fact suggests that there might be a way to separate and identify changes due to either variable—lift-off or conductivity. And there is, but we'll go into that a little later in the program.

At 100 percent IACS a change in lift-off is primarily a change in the horizontal component of the impedance-plane diagram. Note also in Figure 7-7 that at that same point on the chart that a change in conductivity is primarily a change in the vertical component of the impedance. We have been speaking of only one point on the diagram—the 100 percent IACS point. If we plotted the lift-off curve for each material we would get a family of lift-off curves that looks something like Figure 7-8.

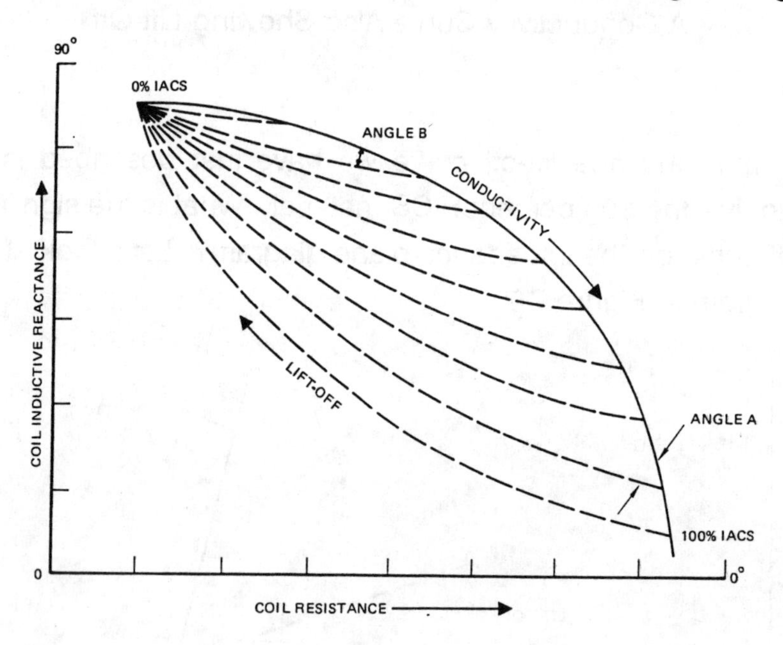

Figure 7-8
Changing the Included Angle Between Conductivity and Lift-Off

Now let's compare the angle at which the lift-off curve approaches the conductivity curve. Compare the angle marked A with the angle marked

B. It would be easier to differentiate between conductivity changes and lift-off changes at the lower end of the curve where the difference in direction of the two variables is greatest. It would seem, then, that in testing materials that have less conductivity it would be more difficult to differentiate between the variables of conductivity and lift-off. And so it is, but there is a way to improve that situation.

Do you recall that we said that the faster the magnetic field changes in a conductive material the greater the eddy current induction? Since this is so, then increasing the frequency of the alternating current through the coil will cause more current to flow in the material. Increasing the frequency also increases the inductive reactance component of the impedance.

All of these factors that are affected by a change in the operating frequency have an effect on the impedance-plane diagram. First, the change in inductive reactance changes the impedance values so we will have to plot a new curve for each new frequency.

When we plot the new curves we discover that in spite of the values being different the curves are similar in many ways. One of the differences is that the materials are moved along the curve towards the lower end of the curve. The operator must continually keep in mind that when he changes the frequency he also changes the depth of penetration—higher frequencies have less depth of penetration.

Figure 7-9 shows the impedance-plane diagrams plotted for a low frequency (20 kHz), a medium frequency (100 kHz) and a high frequency (1 MHz).

Note that the position (locus) of any point representing a given material shifts towards the lower end of the curve as the frequency increases. Thus, it is possible to improve the ability to separate the two variables of conductivity and lift-off by increasing the test frequency.

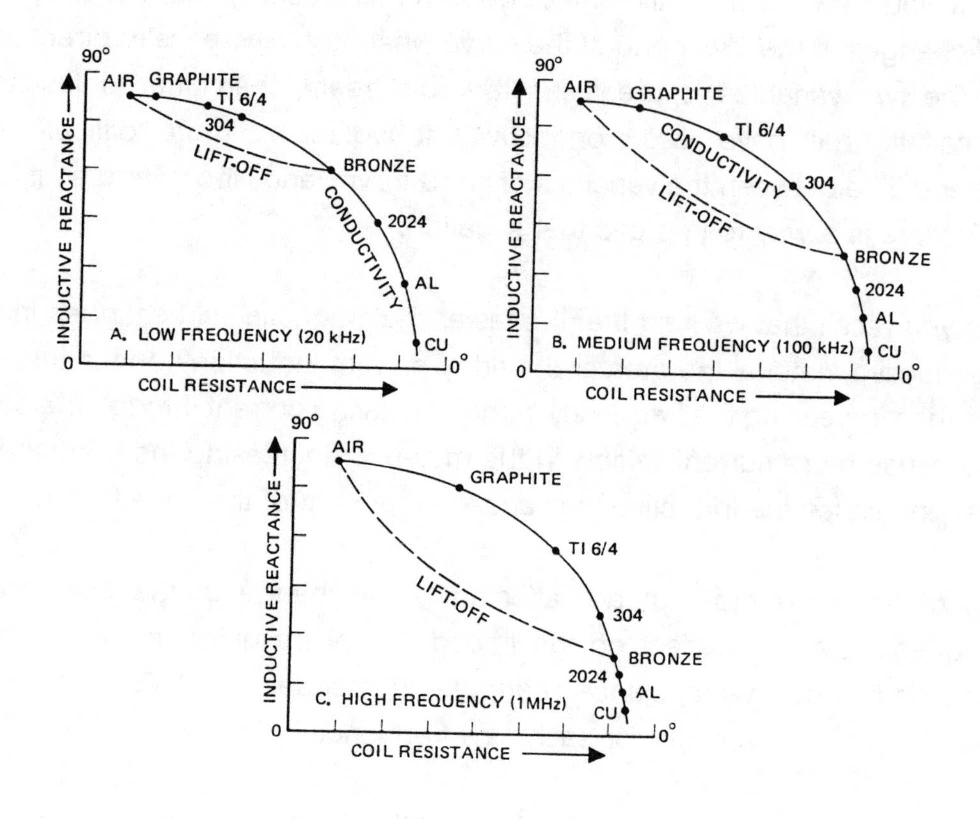

Figure 7-9
Changes in the Conductivity Curve as Frequency Changes

So far in this chapter we have shown the effects of changes in conductivity, lift-off, and frequency on the impedance-plane diagram. Another factor that affects the change of impedance is the dimensional factor of materials that are thinner than the depth of penetration.

To show the effect of changes in dimension on the impedance-plane diagram, we measure the impedance of the test coil as it is placed on varying thicknesses of the material under test and plot these points on the impedance-plane diagram. Figure 7-10 shows the results of such a test on varying thicknesses of brass at a frequency of 120 kHz.

The thickness curve meets the conductivity curve for the first time at the point where the thickness equals the standard depth of penetration of the eddy currents. Increasing the thickness of the brass beyond that point has little effect on the impedance of the test circuit.

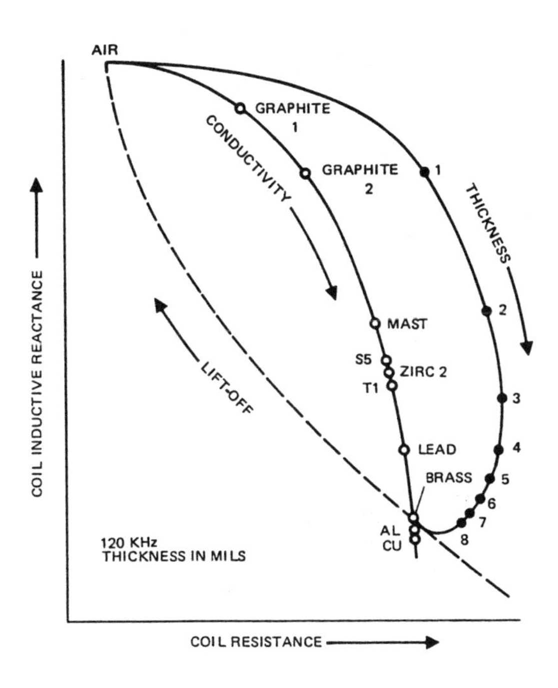

Figure 7-10

Adding Thickness Variations to the Conductivity/Lift-Off Curve

You will recall that the standard depth of penetration depends upon the conductivity of the material and the frequency applied to the test coil. Figure 7-11 shows an expanded view of the impedance-plane diagram showing how the thickness variations differ for three different kinds of material (the frequency is held constant). The numbers along the thickness curves represent the thickness of the material in mils (thousandths of an inch). Based on this graph, notice that the effective depth of penetration in the brass is about 40 mils (1 mm) and in the lead about 65 mils (1.7 mm).

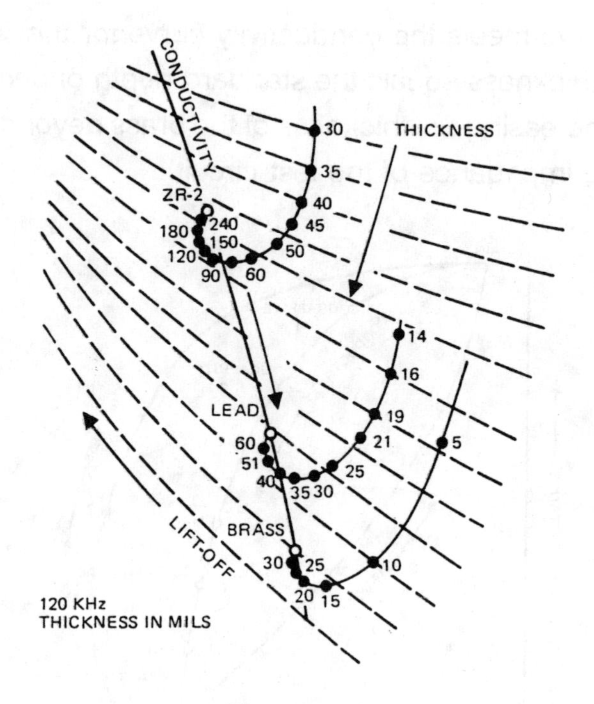

Figure 7-11
Effects of Thickness Variations

Thus, from the impedance-plane diagram, we can see that the depth of penetration in lead is greater than the depth of penetration in brass. Remember that as we move upwards along the conductivity curve the conductivity of the different materials is decreasing. In a previous chapter, we said that the depth of penetration is affected by the conductivity of the material. The impedance-plane diagram bears this out.

Also note in Figure 7-11 that the conductivity is increasing as we move down the conductivity curve. Note also the thickness of the ZR-2, the lead, and the brass at the points where the individual thickness curves meet the conductivity curve. These thicknesses correspond to the depth of penetration of the eddy currents into the individual materials. Note that these key thicknesses are decreasing as the conductivity increases. Thus, depth of penetration decreases as the conductivity increases. The impedance-plane diagram bears out those facts that we explained in a previous chapter.

There is one more thing to note about the thickness curve shown earlier in Figure 7-10. Notice the spread between 7 and 8 mils on the thickness curve. Compare this spread with the spread between 1 and 2 mils. The eddy currents are more sensitive to dimensional changes when the material is thinner. This is plainly shown on the impedance-plane diagram. Notice the change in impedance (the space) between the thickness of 1 mil and 2 mils on the thickness curve. Compare this change in impedance with the change in impedance between the thickness of 7 mils and 8 mils. The change is much greater between 1 and 2 mils.

There is one more effect that we want to point out in Figure 7-12. What happens to the thickness curves at different frequencies? The figure on the left shows an impedance-plane diagram plotted with the frequency at 60 kHz; on the right, at 120 kHz.

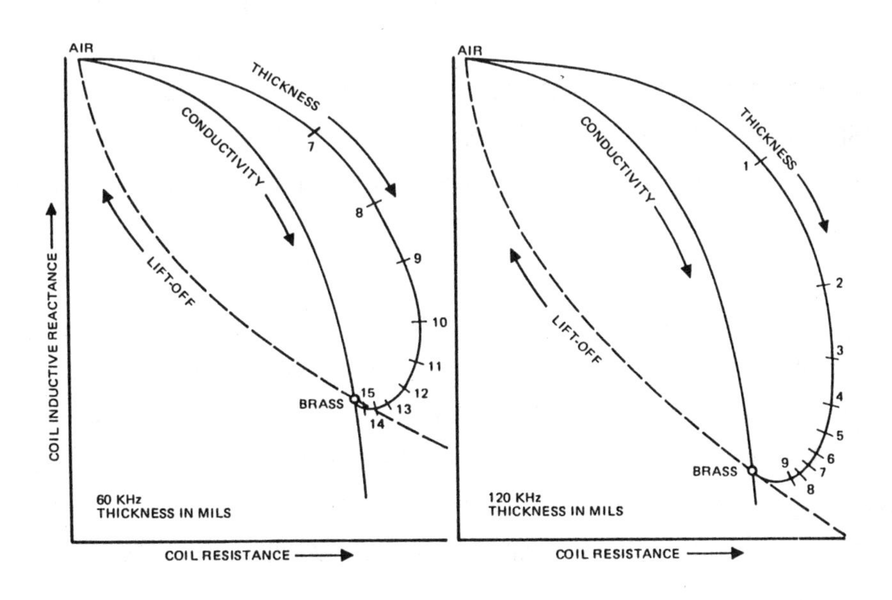

Figure 7-12
Variations in a Thickness Curve as Frequency Changes

- As the frequency increases, the depth of *penetration decreases*. Note that the diagrams showed higher thickness on the thickness curve at the lower frequencies.

- As the frequency increases, the *conductivity increases*. Note that brass is further away from the nonconductive material (air).

- As the frequency increases, the *sensitivity to dimensional changes* at the thickness of 9 mils *decreases*. Compare the spacing between 8 and 9 mils on both diagrams in Figure 7-12. The closer spacing indicates less sensitivity. In this same manner, we would pick any point and show that the sensitivity to thickness changes has decreased as the frequency increases.

Now that you have been thoroughly introduced to the impedance-plane diagram, let' see how the diagram can help us to suppress those variables that are not of interest or that can help us to obtain the best possible results with a minimum of interference. When using the type of equipment where the bridge network may be balanced (nulled) by adjustment of resistance and inductive reactance, the bridge may be set to operate from any point on the impedance plane. Let's see how this capability is used so that the response is more sensitive to one variable and less sensitive to the other.

Suppose that we want to check the conductivity of an alloy, say Alloy A. Let's "dial in" the right amount of inductive reactance and resistance into the balancing impedance so that the equipment is set to operate from the point marked D on the impedance-plane diagram shown in Figure 7-13. If everything is as it should be, the impedance of the inspection coil, when it is placed on the material, will lie on the conductivity line at the point marked "Alloy A."

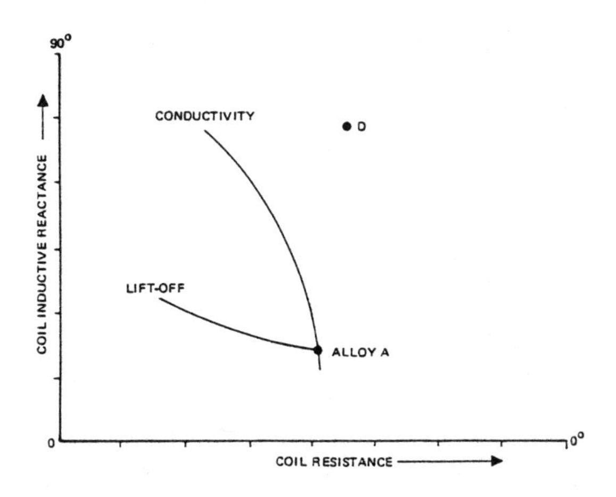

Figure 7-13
Setting Up a Meter-Based Conductivity Test

Suppose now that we want to detect a change in contact and, at the same time, suppress any effect of lift-off. We selected point D as the operating point for this very reason. The diagram in Figure 7-14 below shows why.

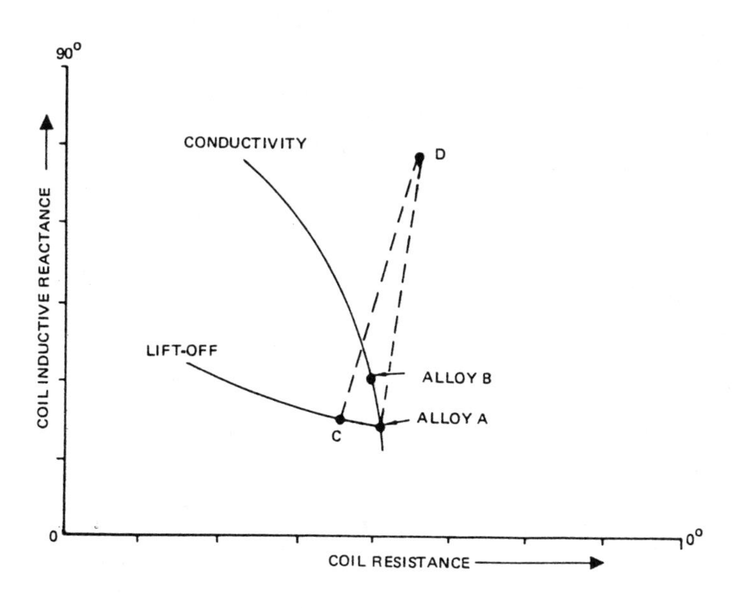

Figure 7-14
Suppressing Lift-Off with a Meter-Based Tester

In shifting our inspection coil from Alloy A to Alloy B, the meter reading will change, since the distance from point D to Alloy A *is not* the same as the distance from point D to Alloy B. The change in the meter reading indicates the difference in conductivity between the two alloys. With point D selected as the operating point, lift-off will not affect the meter reading, since the distance from point D to point C is equal to the distance from point D to Alloy A.

Now we want to point out that point D is not the only point that we could have used as an operating point. We could have used any point that is equidistant from points C and Alloy A. Students of geometry will recognize that the locus of all the points that are equidistant from two points on a plane is the perpendicular bisector of the line drawn between the two points.

For those who are not students of geometry, all this means is that we find a point midway between Alloy A and point C and draw a perpendicular line at that point. In Figure 7-15, we have drawn that line (D to E) and have labeled it the "LIFT-OFF SUPPRESSION LINE."

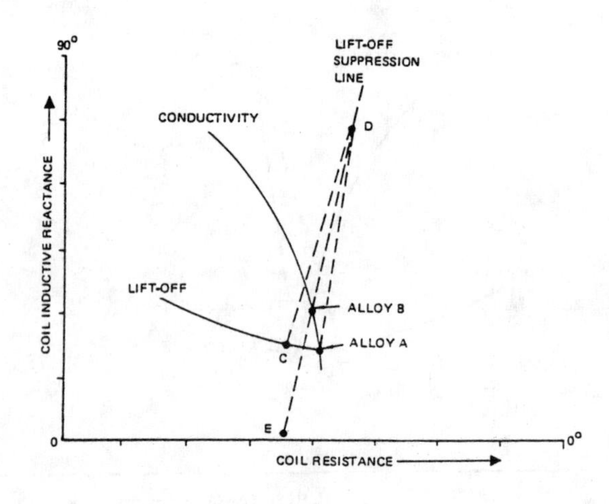

Figure 7-15
Suppressing Lift-Off with a Meter-Based Tester

Point D in Figure 7-15 was selected as an operating point because it was equidistant from two points on the lift-off curve—point C and Alloy A. Point E is also equidistant from the same two points, so it also could have been used as an operating point.

We also want to make sure that you understand that we could use any point that lies on the line E-D since any point on that line is equidistant from point C and Alloy A. If the operating point selected is too far away from the conductivity curve, it is possible that the meter will be driven off scale, but selecting an operating point considerably away from the conductivity curve will produce the best overall results.

If point D in Figure 7-15 is selected as the operating point and the test probe is moved from Alloy A to Alloy B, the meter reading will decrease. If point E is selected as the operating point, the meter reading will increase as the probe is moved from Alloy A to Alloy B. It is possible that better sensitivity will be obtained using an operating point located on one side rather than on the other. In practice, the best procedure is to evaluate operating points on both sides then choose whichever gives the best results.

Now that you have seen how a particular operating point can be selected so that variations in lift-off are suppressed, you probably realize that in the same fashion an operating point can be selected that will suppress variations in the conductivity variable. In measuring the thickness of nonconductive coatings on conductive materials, it is required that the lift-off variable be measured and that the conductivity variable be suppressed. Figure 7-16 shows how the conductivity variable is suppressed by selecting the proper operating point.

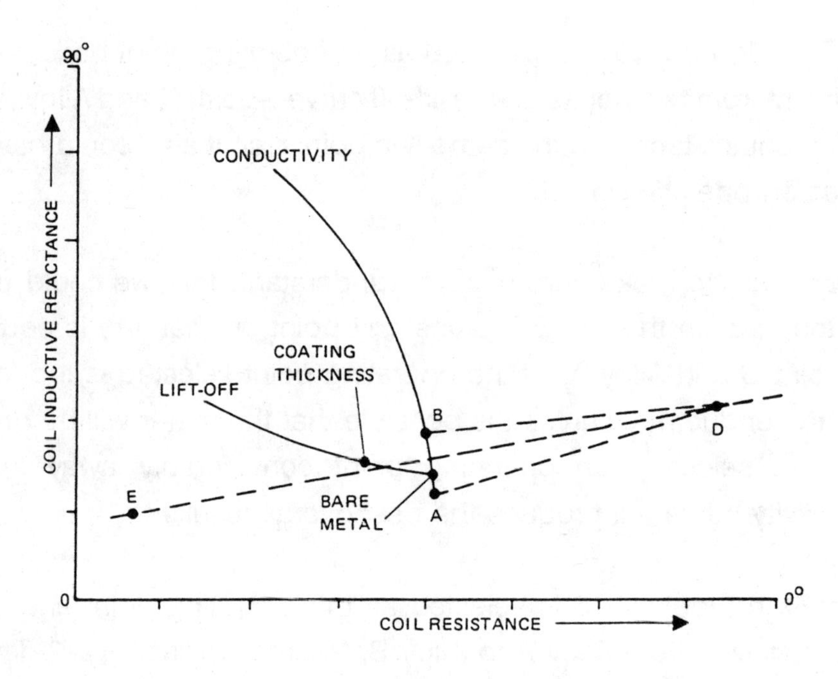

Figure 7-16
Measuring Lift-Off with a Meter-Based Tester

It is important to note that the dashed line from D to E in Figure 7-16 is the conductivity suppression line. The selection of an operating point that lies somewhere on that line will cause the conductivity variable to be suppressed without affecting the indication of a change in lift-off.

Also note that the conductivity suppression line in Figure 7-16 is essentially perpendicular to the conductivity curve at the point of interest. The considerations for the final selection of the operating point are the same as those we gave for the final selection of the operating point when suppressing lift-off. In short, select the point that gives the best results.

When a cathode-ray tube (CRT) is provided as part of the test equipment, it is possible to set up the equipment so that the impedance of the test coil will cause a dot to appear on the face of the CRT. The position of the dot is determined by the test coil impedance.

Now if we take a sample of several different kinds of conductive materials and test each one with the system we just described, we will find that the dot on the tube will appear in a different position as we test each type of material. If we note the position of each dot as it appears on the screen, we will find we have a familiar-looking curve. Since we were taking readings on several different types of materials, the curve has to be the conductivity curve.

We can also plot the lift-off curve by taking readings of varying thicknesses of nonconductive material placed on top of the material that we are about to test. In this way, the entire impedance-plane diagram can be plotted. During actual testing of specimens, the impedance of the test coil will cause a dot to appear at some point on the screen. Its position with respect to the impedance-plane diagram tells the operator what has occurred within the specimen.

By testing reference standards, the range of allowable discontinuities may also be identified on the CRT. The reading obtained on a crack in a standard, for example, can be set at a given point on the screen. In this manner, all the acceptable "parameters" can be located in one part of the screen to immediately show good material from material which shows unacceptable signal response.

An advantage of using a CRT is its extreme flexibility. For example, the equipment may be set up so that the display is rotated, as shown in Figure 7-17.

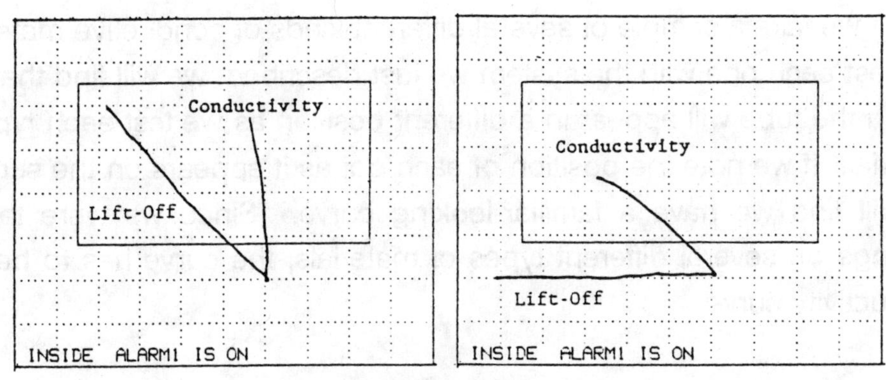

Figure 7-17
Advantages of a CRT (X-Y) Display

The normalized impedance curves that we have shown so far in this chapter have had what we might describe as "absolute" responses. The changes that we see graphed (conductivity, permeability, lift-off, etc.) in each set of curves is a line based on a set of points in response to a test or material change.

As we can see in the sets of curves in Figure 7-17, the major disadvantage of eddy current testing is that it responds to an extremely large range of variables. It is sometimes difficult to decide what, if anything, the output signals seen have to do with the material we are inspecting.

In Figure 7-17, we saw that it is possible to rotate these normalized impedance curves on the screen of an X-Y impedance-plane device (CRT) so that certain variables are at known positions on the screen. A signal on the screen *may* indicate major damage to the material; however, an identical signal with the same size and direction might have absolutely nothing to do with the material. When we detect an output signal that has nothing to do with a discontinuity or material changes, then we classify it as a "noise signal."

Our task in calibrating and interpreting EC equipment is to be able to separate the "signals of interest" (S) from the different "noise" (N) sources. In this way, we can develop a method for assuring that we have good flaw detection capability. The classic signal-to-noise (S/N) ratio that is said to provide this capability is 3 to 1. We quite often have to live with much lower levels to detect minimum flaw sizes required.

The decision of how to adjust the CRT display is the most important step in setting up a good EC examination. Let's take a few minutes and discuss the CRT-type presentations and how we decide to rotate each of the signal patterns to help make a test easier.

Let's try something. Why don't we create a table that shows how a clock's minute hand moves. We will look at both the change in time and the change in position of the minute hand. Let's start with what we know.

- We know that there are 60 minutes in an hour (one full sweep of the clock face).
- We know that one hour will bring the minute hand all the way through a complete circle.
- We know that a complete circle has 360° in it.
- If we divide the 360° of the clock face by the 60 minutes on the clock face, that tells us that each minute on the clock's face is equal to:

  $360° \div 60$ (minutes) $= 6°$ per minute.

  Each 5-minute interval could be expressed as:

  5 minutes x 6° per minute $= 30°$

We can measure the change in the minute hand position by measuring the number of degrees between the 12 o'clock position and wherever the minute hand happens to be at any point in time. It would look something like Figure 7-18.

CLOCK	PHASE	TIME
12:00 position	0°	
12:05 position	30°	5 minutes
12:10 position	60°	10 minutes
12:15 position	90°	15 minutes

Figure 7-18
Relationship to Time and Phase Angle

It is apparent that the angle between the minute hand and the 12:00 position increases as time increases, or:

AS TIME INCREASES, THE PHASE ANGLE INCREASES

Our statements about phase and time in Chapter 1 with respect to electromagnetic properties can be just as easily applied to our discussion of the CRT presentations showing responses to material changes.

Remember in Chapter 2 we said that eddy currents do not happen instantly throughout a material. They are strongest and occur first very close to the coil. Over some time (based on frequency), the primary magnetic field strength and distribution will continue to grow in response to the increasing current flow through the coil. This will continue to some and then start decreasing again toward zero. But while the primary field strength is growing, it creates eddy currents deeper and deeper in the material over the same time period.

Figure 7-19 is a close-up of a tubing standard looking at a cross section of one tube wall thickness. The coil is inside the tube (ID bobbin) and the discontinuities are machined from the tube OD. Pay close attention to the amount (or thickness) of "good" material in each area. Notice that in this example, there is 0.010 inch (.25 mm) of nominal material between the ID surface and the bottom of the 80 percent flaw.

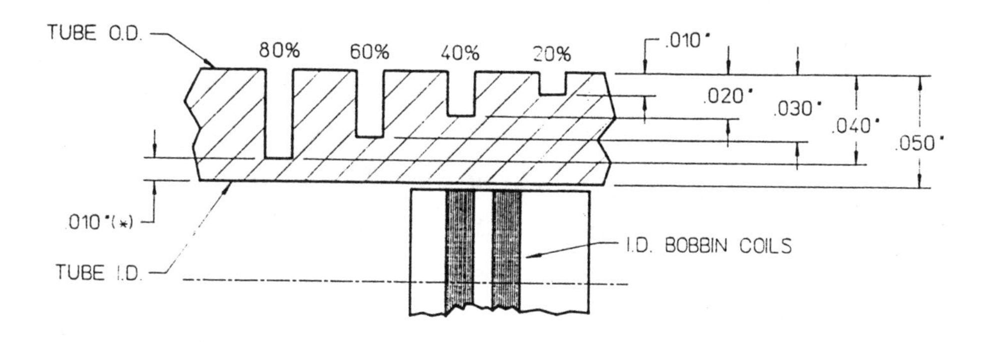

Figure 7-19
Tube Cross Section Showing Material Removed vs. Remaining Wall Thickness for Several OD FBH Flaws

At a given frequency, this thin 0.10 inch (.25 mm) layer will quickly be penetrated by the primary magnetic field (and the eddy current field being generated by it). The 80 percent flaw is considered to be a "localized disruption in conductivity" just above this thin layer of nominal material. The flaw will be detected very quickly, since it is just below the inspection surface. If the 80 percent flaw in Figure 7-19 is sensed very quickly, then the phase angle *difference* between "time zero" (the ID surface) and the point in time where the 80 percent flaw disrupts the eddy current flow is going to be relatively small. It is going to take a relatively longer period of time for our eddy current field to penetrate and respond to the material disruption at the 20 percent flaw level.

When we compare the two eddy current test output signals, we can make one of two statements.

- **The measured phase angle of the 20 percent OD flaw "lags" the 80 percent OD flaw.**

OR

- **The measured phase angle of the 80 percent OD flaw "leads" the 20 percent OD flaw.**

Keep in mind, however, that this relationship is based on the fact that our coil is on the inside or "near" surface and our flaws are on the outside or "far" surface. If we change this situation by using an OD coil and look at the question of leading and lagging responses from ID flaws, the situation would be exactly reversed.

Something that "leads" another event occurs earlier in time. If it happens earlier in time, then the amount of measured phase rotation from some arbitrary starting point will be smaller. The phase angles that we talk about in EC signal calibration and analysis are always measured *from the same starting point.*

CRT-based inspection equipment became common in the eddy current inspection field some years ago. The largest application is in the thin-walled nonferromagnetic tubing market. In recent years, CRT-based EC testers have been in greater demand in the aircraft market due to specifications requiring higher detection and discrimination capabilities.

It was decided early on that the nine o'clock position on a CRT screen would be the point that we call zero degrees (0°).

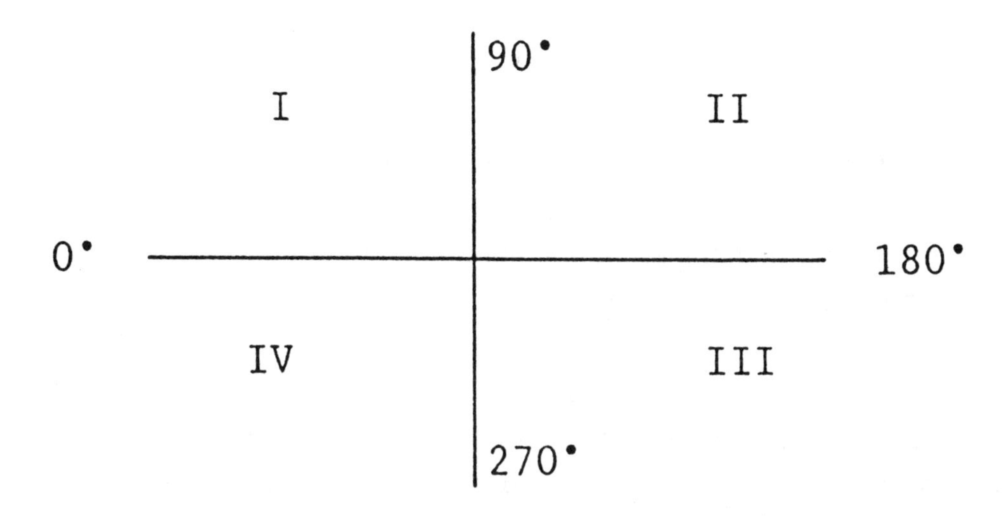

Figure 7-20

Phase Angle Orientation on a CRT (X-Y) Display

For those of you who have had a class in geometry you may notice that this is exactly 180° *out of phase* from what you learned about two dimensional systems. It is not critical that we get into why this decision was made at this point, just that it has become the accepted norm in our little world. If we see a response in zone I of the screen shown in Figure 20, it would be considered "*leading*" *(ahead of) a signal that occurred on the screen in zone II. Measured phase angles of inspection signals occurring in zone I would be less than* the measured phase angles of response signals occurring in zone II.

The first step in setting up CRT-based equipment is to look at the different types of noise sources and determine their phase angle locations on the screen. Anything that is not flaw related could be called a source of noise if it might "mask" or distort actual flaw information. This could include:

- signals from tube wall variations (geometry),
- signals from material variation (permeability),
- inspection/probe noise (lift-off, fill factor),
- signals from external electrical sources (spikes),
- signals from material changes outside the intended inspection zone (support structures, deposits).

The most common approach for adjusting signal rotations on the CRT during the system calibration process is to *put all "noise" sources related to geometry* on the horizontal axis of our screen (0° - 180°). Next we would look at the calibration standard flaw responses and determine their phase angle location on the screen. It may be necessary to fine tune our phase angle and/or amplitude adjustments to assure that we have eliminated most of the noise out of our "detection" range.

If we have carefully selected our test frequencies and made the proper "noise" adjustments, then this is going to leave our other major variable, conductivity shifts (flaws), reacting alone along the vertical axis, as shown in Figure 7-21. If we now monitor the vertical signal component, we should have an improved detection potential with a minimum of distortion.

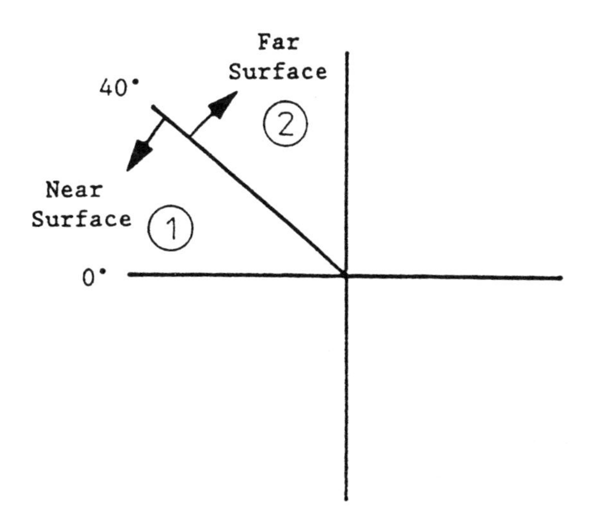

Tubing Inspection: **ID Bobbin Coil**

1. Flaws on the ID surface
2. Flaws on the OD surface

Tubing Inspection: **ID Encircling Coil**

1. Flaws on the ID surface
2. Flaws on the OD surface

Surface Inspection:

1. Flaws open to the inspection surface
2. Flaws below the surface

Figure 7-21
Detecting Signals of Interest on a CRT Display
After Rotating Lift-Off to the Horizon

This "noise" rotation is used for many different applications. In Figure 7-22 on the next page, we can see how noise rotation has changed our detection potential of the flat-bottomed holes seen on a tubing standard. If we look at the two strip-chart presentations we can begin to see what we mean by S/N ratio. If you compare the two strip-chart traces, you will see that they are very different. The upper strip-chart in Figure 7-22 shows the combined result of both flaw and noise information.

When we *rotate noise to the horizontal* position in the lower section, we see that the noise has little or no response into the vertical plane; all of the information left in the vertical channel is flaw related.

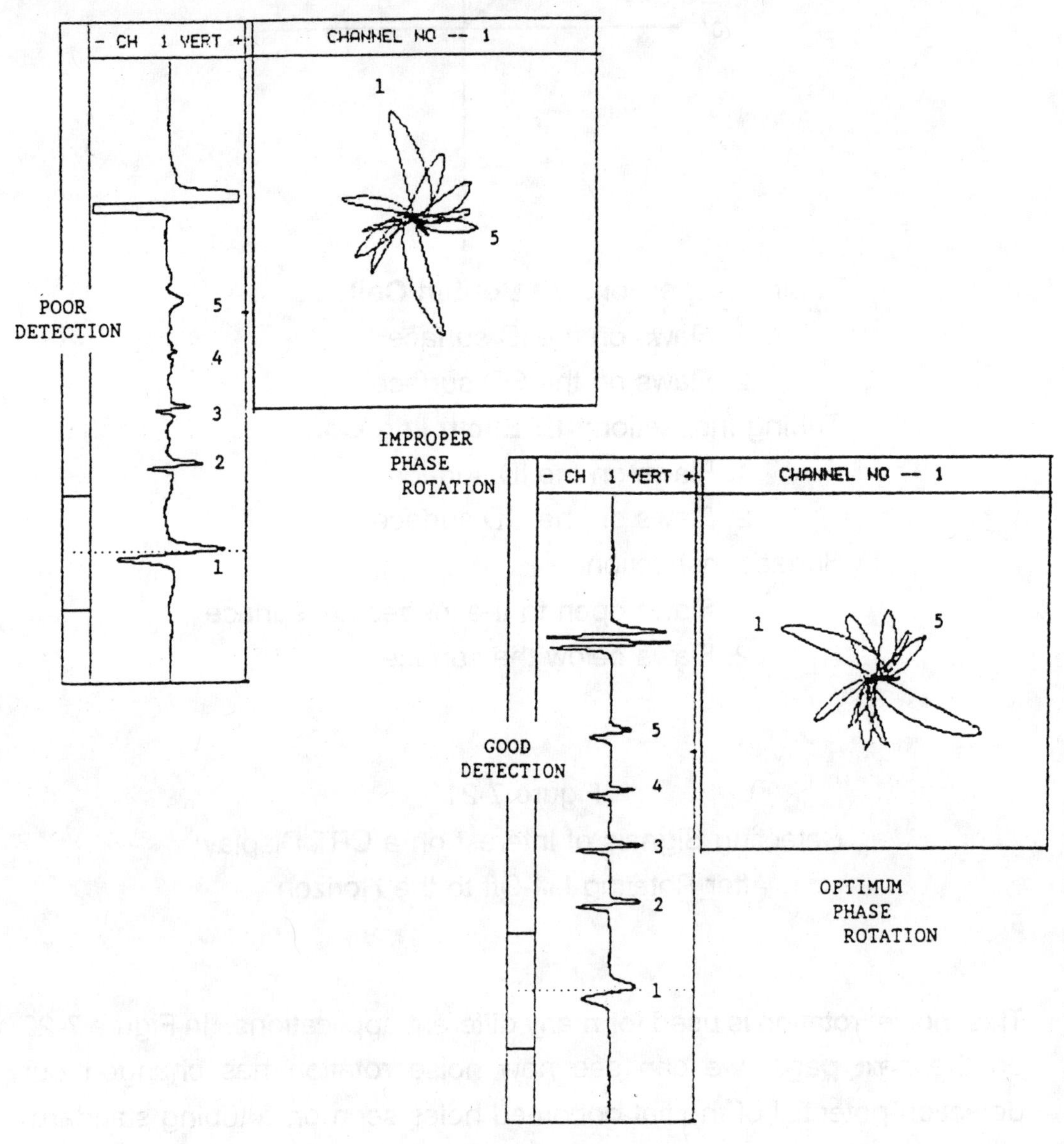

Figure 7-22
Improved Flaw Detection by Applying Phase Rotation

CHAPTER 8

TEST EQUIPMENT

In eddy current testing, many instruments are designed for a particular application to a testing problem. Others are designed for general-purpose use. Within the general-purpose category some are designed to test for a particular variable, while others are designed with the capability of testing for several variables. A detailed explanation of all the designs of eddy current instruments available on the market is beyond the scope of this handbook. We will, however, attempt to explain the operation of two of the more common types.

- Conductivity Testers
- Crack Detectors

The operation of particular types of equipment produced by different manufacturers for the same purpose are quite similar, even though the controls may bear different names. For specific operating instructions for a particular instrument, the operator must refer to the manufacturer's handbook for that instrument.

Conductivity Testers

Conductivity testers are defined as simple instruments designed only to check the conductivity of various types of materials and their alloys. For example, a conductivity tester cannot be used to detect discontinuities. It lacks any means to adjust the frequency supplied to the coil to provide for variations in the depth of penetration. For the same reason, it cannot be used to detect thickness changes or suppress lift-off.

Conductivity readings will be affected by lift-off. In using a conductivity tester the operator must be continuously aware of all the factors that can affect conductivity before reaching any conclusions based on the output signals seen. Figure 8-1 shows one type of conductivity tester.

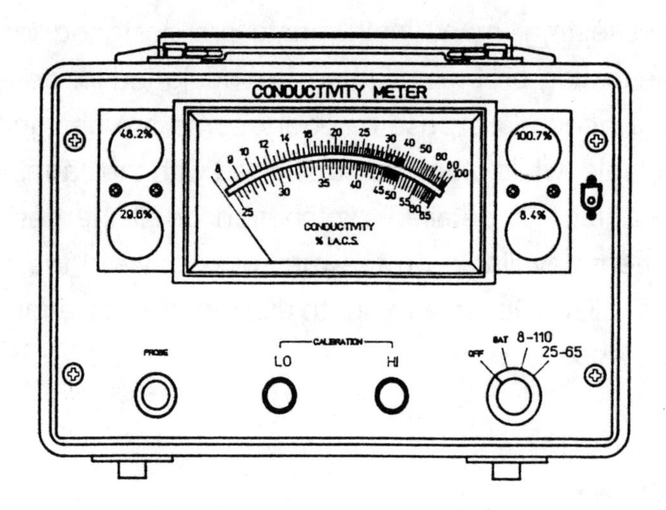

Figure 8-1
Meter-Based Conductivity Tester

The tester is supplied with several blocks of known conductivity that are used to calibrate the instrument prior to its use. Each standard has its conductivity in percent IACS stamped on it. The first step in the calibration of this instrument consists of selecting the expected conductivity range to be used during the test. The options are a high range (8 percent to 100 percent IACS) or a mid-range which is more centralized around the aluminum family (25 percent to 65 percent IACS).

- Set the probe on the *lowest* standard available within the selected range and adjust the low scale reading with the LO calibration knob.

- Set the probe on the *highest* standard available within the selected range and adjust the high scale reading with the HI calibration knob.

Verify the precision of the readings over the range you selected with additional calibration standards within that range. If necessary, fine tune the final HI and LO calibration adjustments to meet the specification being used. Once the instrument is calibrated, the HI and LO controls are not touched during the ensuing tests.

Tests are conducted by placing the test coil firmly on the test specimen. The reading on the meter is the conductivity of the specimen. Most conductivity testers have some arrangement to calibrate the instrument that is similar to the one we've explained here.

There are two major differences in most of the instruments considered to be conductivity testers.

- The first is in the signal generation section, either fixed or variable frequency.

- The second is the style of output, either analog or digital.

By using fixed frequency devices, you know what thickness of material should be inspected because you will also know the eddy current standard depth of penetration at that frequency in the material being tested.

The term "variable frequency" in conductivity testers is a little misleading. In some models you cannot actually choose the frequency you want. The instrument will change operating frequencies internally during the calibration process, depending on the material and thickness being used.

Instead of having an accepted working thickness range in a given material, these instruments shift their own frequency to some point where a reading is obtained that is not controlled by the actual thickness. This is normally used to get a reading in very thin materials.

The second major variation is in the instrument output mode. The options are either analog or digital. The meter-reading instrument in Figure 8-1 is an example of an analog device. Digital output devices might include a liquid crystal diode (LCD) numerical display instead of a meter.

One of the major areas for applications of conductivity tests is the aircraft market. They are used to verify material content (alloy), as well as material temper (hardness) based on heat-treat processes used. These tests are performed on all material to be used in aircraft construction. The larger aircraft manufacturers have done a lot of testing and comparison of many different styles and models of conductivity testers. The accuracy and precision of each conductivity tester is known and accepted within certain limits. Customer codes and specifications defining material acceptability and technique guidelines should be considered prior to testing.

Crack Detectors

Crack or discontinuity detectors are more complex instruments than conductivity testers. The major advantage of these more complex systems is that it is now possible to choose the frequency, *or frequencies*, that will be used. This gives the operator the ability to adjust the depth of penetration to meet the needs of each test.

Single-frequency equipment may be dedicated to one type of task, such as an on-line inspection process. These instruments may use a meter-based output or have a CRT screen presentation. Although the frequency is variable in the instrument in Figure 8-2, it is capable of only using one frequency at a time. It uses what is sometimes referred to as an "uncalibrated" meter approach. You should remember that our conductivity meter read-out was "calibrated." It showed a number (in percent IACS) that directly related to a materials conductivity.

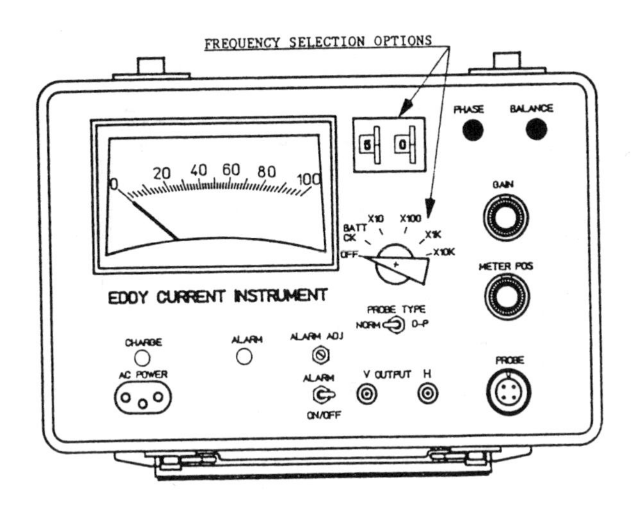

Figure 8-2
Single-Frequency Crack Detector

Of the two meters shown in Figure 8-3, the one on top (1) is "calibrated." There is a number shown on the scale that means something to the eddy current operator. The lower meter display (2) is "uncalibrated." We cannot directly interpret a percent of material change based on a needle position on the scale that we have available. We can tell if the needle goes "up-scale" or "down-scale," we can tell if the change is "large" or "small" based on the amount of needle deflection, but we cannot read the scale directly.

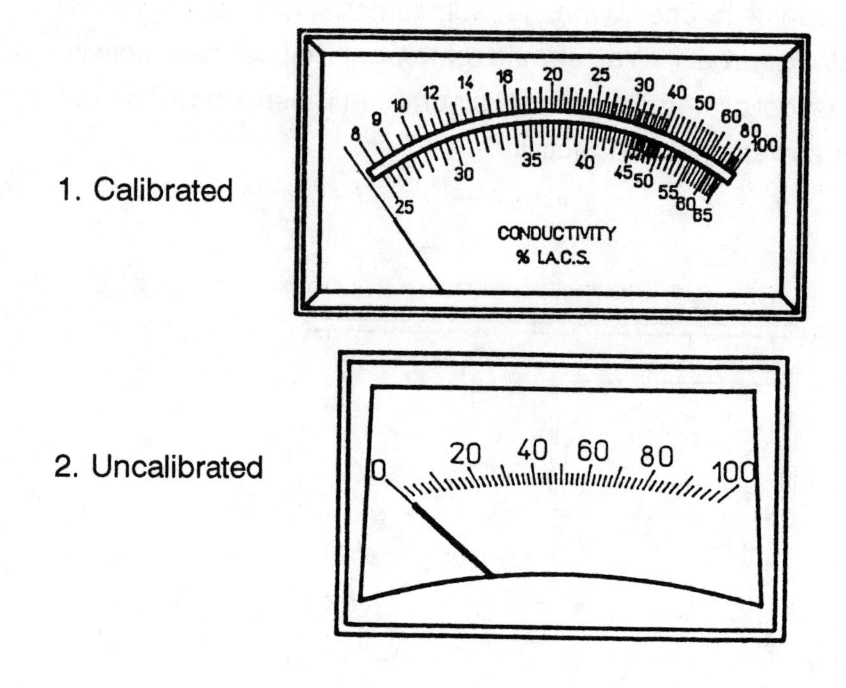

Figure 8-3
Comparison of Calibrated and Uncalibrated Meters

Of course, the "primary side" of our equipment must go through periodic maintenance and repair (**calibration**) by a qualified electronics technician. The "secondary side," which includes the output screen or device that we are using to monitor our material condition, does not necessarily have to be "calibrated." Because we are limited in our interpretation of this type of output, we must be very careful in the setup of these devices. The standards used become very critical. We have to control as many aspects of the test as possible to make sure that none of them give us a false defect reading which would lead to the rejection of good parts.

As test conditions present more and more variables to us, it may not be possible to rely on meter-based testing. On a meter-based device, we are only looking at one-half of the information available to us. If we plotted the voltage shift being sensed by our bridge circuit, it would be varying in both magnitude (X) and phase (Y). The signal seen has a magnitude, but in what direction? Remember that our test *could be* sensitive to changes in:

Conductivity/Permeability/Geometry

If we can remember that our test may be sensitive to more than one type of material condition under a given set of conditions *or a combination of material conditions*, then it will make our task of understanding those output signals that much easier.

Figure 8-4 illustrates that it may be possible to separate responses that result from either a permeability or a conductivity shift. Perhaps one is acceptable while the other is rejectable. Separating the responses from these two changes may require specialized test probes and techniques, or we may just have to live with the combined signal response.

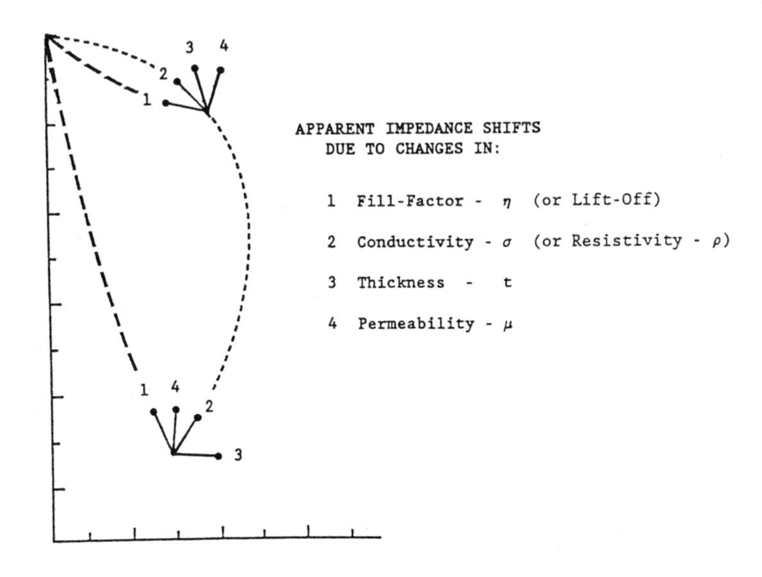

Figure 8-4
Optimizing Test Conditions to Detect Signals of Interest

An aircraft situation where this may occur is a cracks-under-fastener (CUF) exam. The fasteners may be ferritic and have a permeability value of much greater than one. The aircraft skin layers are typically aluminum (nonferritic) and are assigned a relative permeability of one. If you are looking for cracks (localized conductivity shifts) coming out of the fastener holes, then we may expect to have a signal change that is due to the combined material (permeability) and flaw (conductivity) responses.

It is possible that we might be able to better understand the output signal from the test if we look at all of the information provided on the X-Y screen or CRT. This would allow us to measure *both amplitude* (how big [in volts] the change was) *and phase* (what direction the change occurred in) of the response. By using both the vertical and horizontal display voltages, we might very easily make distinctions between those variables that are going to affect our output signal.

The instrument shown in Figure 8-5 has a meter-based output mode. This meter is responding to the voltage shift represented on the vertical axis of our impedance plane. If we take the total signal shift available out through the BNC connectors on the front panel and display it on an X-Y screen (CRT), then we can take advantage of the total package of information. The meter-based devices are capable of displaying information along one axis. The setup process (frequency selection and signal rotation) defines what types of changes we are going to be sensitive to (conductivity, permeability, or geometry) along this axis.

The CRT-based devices are capable of displaying twice as much information. They can provide a more complete picture of the signal changes that we are detecting. Maybe a simple Go/No-Go test is all that is really required. If this can be done well with a meter-based instrument (that meets the customer's specifications), then the problem is solved.

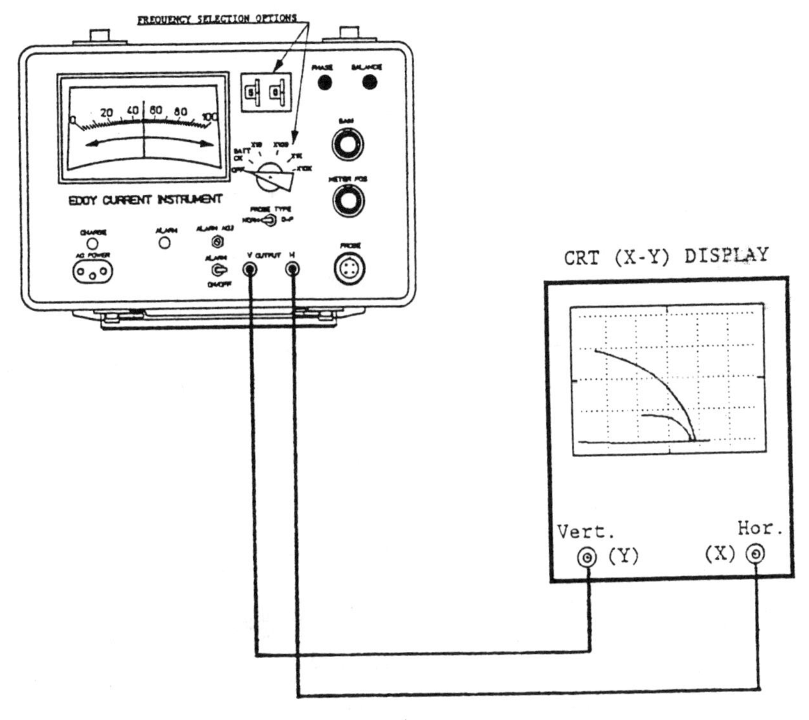

Figure 8-5
Single-Frequency Tester with BNC Outputs
Hooked Up to an External X-Y CRT Display

If the customer is looking for a more complete examination, one that provides him a clearer picture of material characterization and flaw pattern recognition, then the CRT-based options will probably have to be brought into play. So what if the customer is looking to provide a product which has undergone a more comprehensive EC examination process? What other types of equipment options do we have available to us? The next step forward in system capability would be to combine both components into one CRT-based device. One possible option is shown in Figure 8-6.

This unit happens to use what is more accurately defined as a digital CRT. The older versions of test instrument CRTs relied on a focused beam of electrons being emitted by a "gun," much like a television set. This beam would hit the back of the screen and cause phosphorescent materials to be activated to a higher energy level. They would "shine" for a period of time that was controllable. These analog CRTs are being phased out by many of the major suppliers.

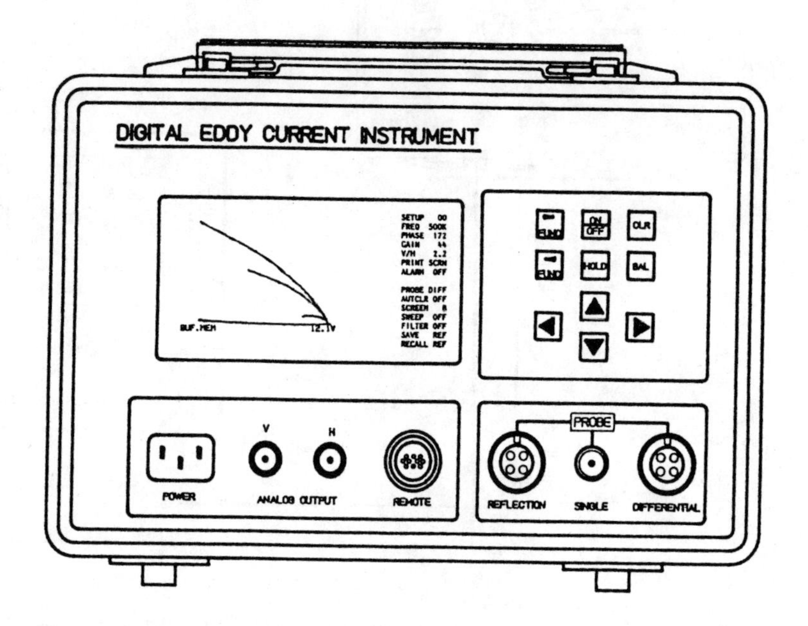

Figure 8-6
Single-Frequency Eddy Current Device with CRT Display

The digital CRTs act much like a computer screen. The screen is divided into "pixels." Each pixel is a small fraction of the entire screen surface area. Each of these zones is turned on or off by the output circuitry. When it is "on," it can be assigned various values concerning either brightness or color.

This is just one change that is affecting equipment production at present. On-board computers and other memory options allow EC systems to be much broader in their range of functions and applications and yet easier to use than some of their earlier analog versions. As technology rapidly grows and expands, the cost for each previous level of technology will become less. A hand-held calculator that costs ten dollars today may have cost hundreds of dollars a few years ago to get the same range of functions. This type of growth has impacted the EC equipment manufacturing market as well.

Some of the other major advantages of the digital CRT presentation are:

- With digital CRTs it is possible to display all of the important information about the system operation and variables right on the screen.

- It is also possible to create a digital box right on the screen. This box, or "gate," can be set so that as long as no signal enters or leaves the box, then nothing happens. The system is balanced so that the dot stays in the center of the screen, just like you would expect for good material. But if the eddy current field is disrupted by a flaw, the response signal will come up and go outside the electronic "gate." The threshold or alarm point for this gate will be "triggered" and can now be used to turn on other alarms, either audible or visual. The nice thing about this is that now the operator doesn't have to be watching the screen and can be watching the probe and inspection piece more closely. The operator can do a very careful scan close to the edge of the test piece and the equipment will indicate when something has changed. The digital CRT shown in Figure 8-7 makes an excellent flaw detection device.

Many of the single-frequency CRT devices have been designed for portability. A few have self-contained battery packs so they can be used without direct access to an AC power source. This makes them ideal for many field situations. Most have a very wide frequency range and some can be used with a wide range of probe types. This makes them applicable in almost any kind of surface scanning inspection for the detection of discontinuities on the near surface as well as the back surface.

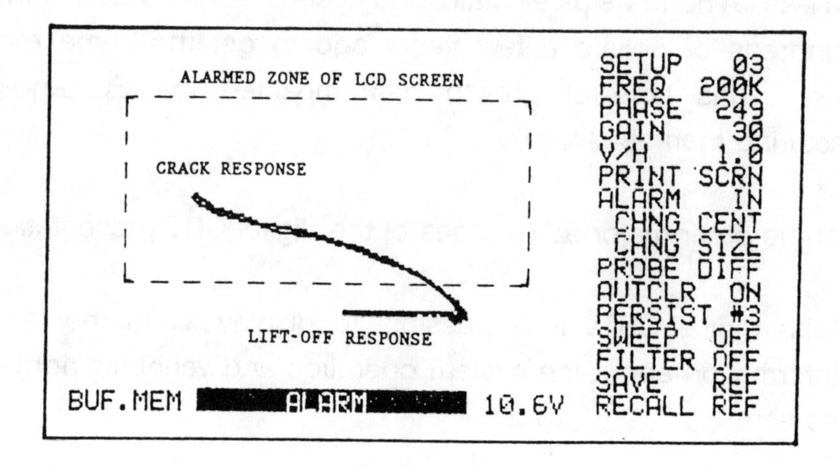

Figure 8-7
Digital CRT Display Options

Even though they have the capability of working over a broad frequency range, they are still limited to one frequency at a time. This means they are going to have limited applications in some types of inspections.

Many tube mills still use single-frequency testers due to the fact that they are inspecting virgin material. The tubing is still fairly clean just after production compared to what could happen to it after it goes into service. The encircling coil technology generally used may do an adequate job of detecting the majority of possible flaw mechanisms.

Advanced Test Equipment

To support growths in population, we must meet growing demands for electrical power. Today's work force relies on advanced electronics and climate control. Many heat exchangers have come into use to support these demands: steam generators, feed water heaters, condensers, and air conditioners to name a few. We must find ways to inspect this large array of tubing and to assure safe and economical operation. The tubing inspection business just got a little more complicated.

Some of these heat exchangers have thousands of tubes. It was soon found that several tests at different frequencies may be required to fully understand what might be happening to a tube. This makes for a very time-consuming inspection. To provide these more complex examinations, two types of equipment capabilities are usually required—multifrequency and multiparameter.

- **Multifrequency**

  This equipment must be capable of supplying more than one frequency to one or more coils. Multifrequency testing has many applications in the aircraft inspection and surface scanning testing areas as well as heat exchanger inspection. Multifrequency testing can provide a more efficient way of collecting enough information about complex systems to make timely and realistic decisions about their condition.

  Figure 8-8 shows the application of more than one frequency to layers of different material (or varying thicknesses of the same material). We can localize our inspection sensitivity to give us more information about compound structures and materials.

  A multifrequency system can have each "channel" set so it will only detect changes to a certain depth in a material. Each channel can be tuned to provide information on smaller and smaller thicknesses of material by simply increasing the frequency within the limitations of the coil being used.

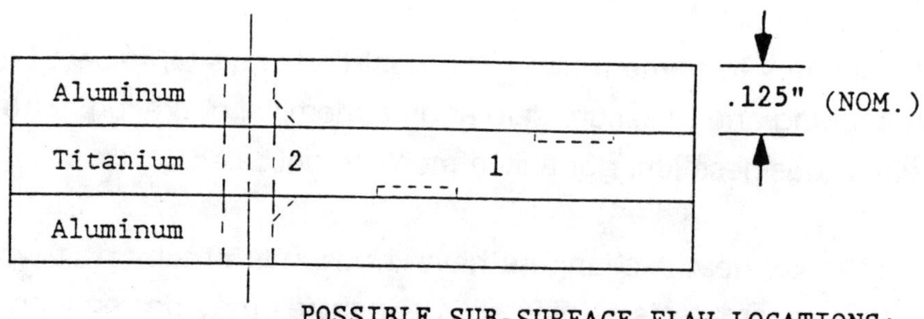

POSSIBLE SUB-SURFACE FLAW LOCATIONS:

1. CORROSION AT INTERFACE
2. STRESS CRACKS IN BOLT HOLES

Figure 8-8
Stacked Materials of Varying Conductivity

Another approach might be where an array of coils is needed to test a complex geometry such as turbine rotors, blades, or wheel hub radii, as shown in Figure 8-9. Each channel could be used to drive a separate coil on a multicoil probe. This might help reduce signal losses due to "ganging" several coils in a series on the same output. This would be less of a load on the circuit and allow each coil to operate independently and with greater sensitivity.

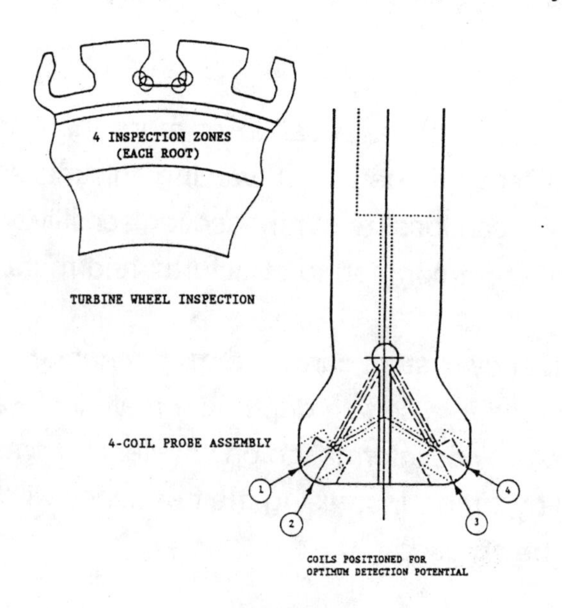

Figure 8-9
Multicoil Array for Testing Complex Geometry

- **Multiparameter**

The equipment must be capable of adjusting and rotating output signals. It may have options for combining or "mixing" these signals in such a way that "noise" or other information about nonflaw conditions can be eliminated from the EC output data.

The *multiparameter* concept of "mixing" two or more data channels is actually quite simple. By using more than one frequency, we can have more than one "picture" of the same point in time in our data. Let's look at a tube-support plate (TSP) intersection signal as shown in Figure 8-10.

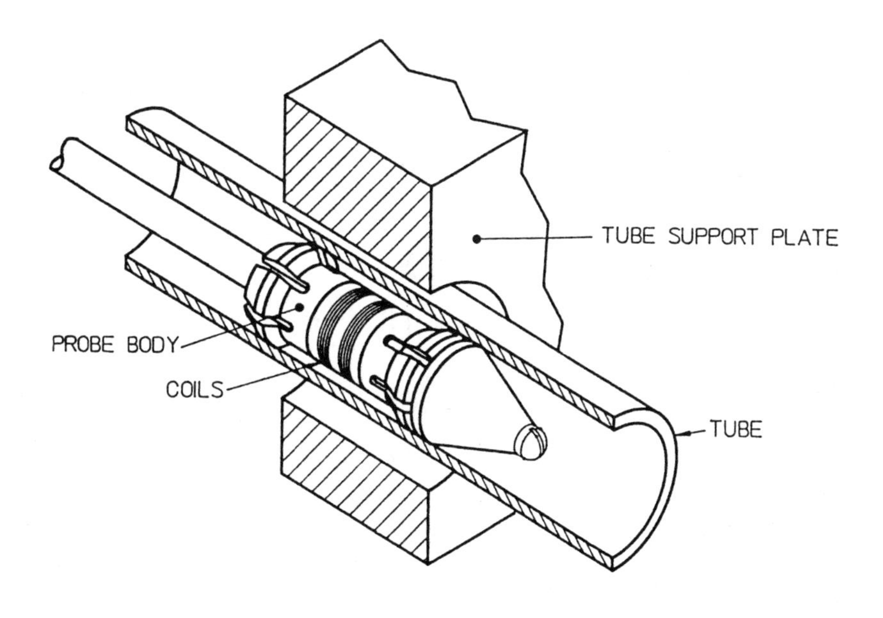

Figure 8-10
One Tube (in Cross Section) passing Through a Tube Support Plate Structure

Our eddy current test is going to be sensitive to all changes in coil impedance. Keep in mind that higher frequencies would not "see" or sense this support plate as well as low frequencies because their standard depth of penetration will be less.

As shown in Figure 8-11, higher frequencies are more sensitive to events or changes in the tube wall than things outside the tube wall (the probe is inside the tube).

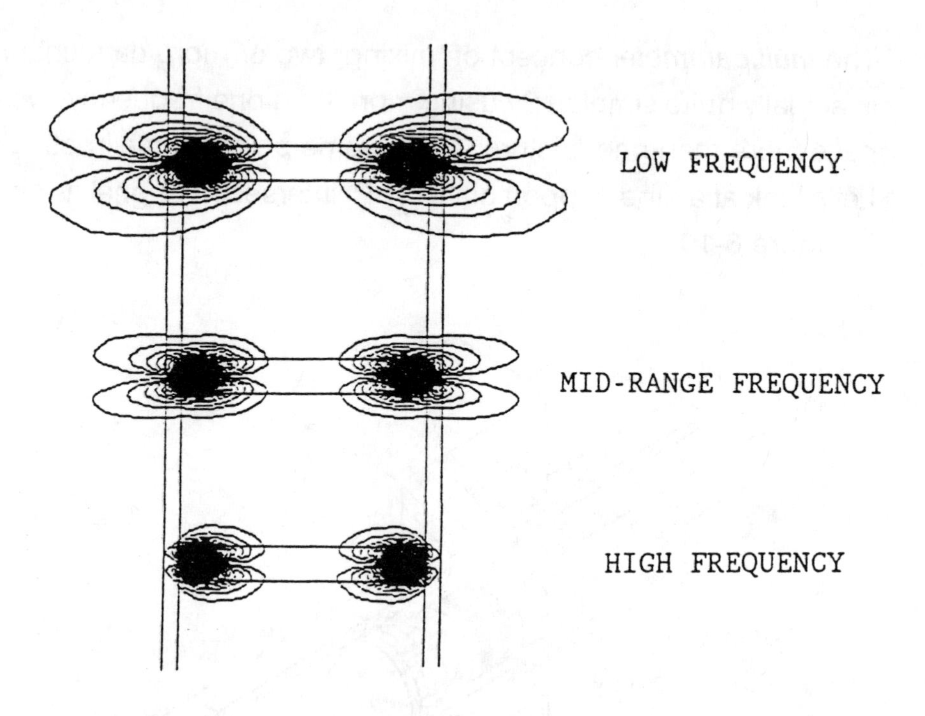

Figure 8-11
Frequency versus Sensitivity

The eddy currents generated by high frequency are going to be weaker than those generated by a low frequency signal by the time they get all the way through the tube wall and begin to sense changes outside the tube wall.

Figure 8-12 shows two possible responses to the detection of an **undamaged** tube support plate (TSP) from Channel "A" (a higher frequency) and Channel "B" (a lower frequency). Both of these channels have been calibrated by adjusting their gain and phase responses to an ASME standard.

You can see that both channels are sensitive to the tube support plate. Unfortunately, many of the possible flaws that might occur are also going to be found in the tube as it passes through or near one of these support structures.

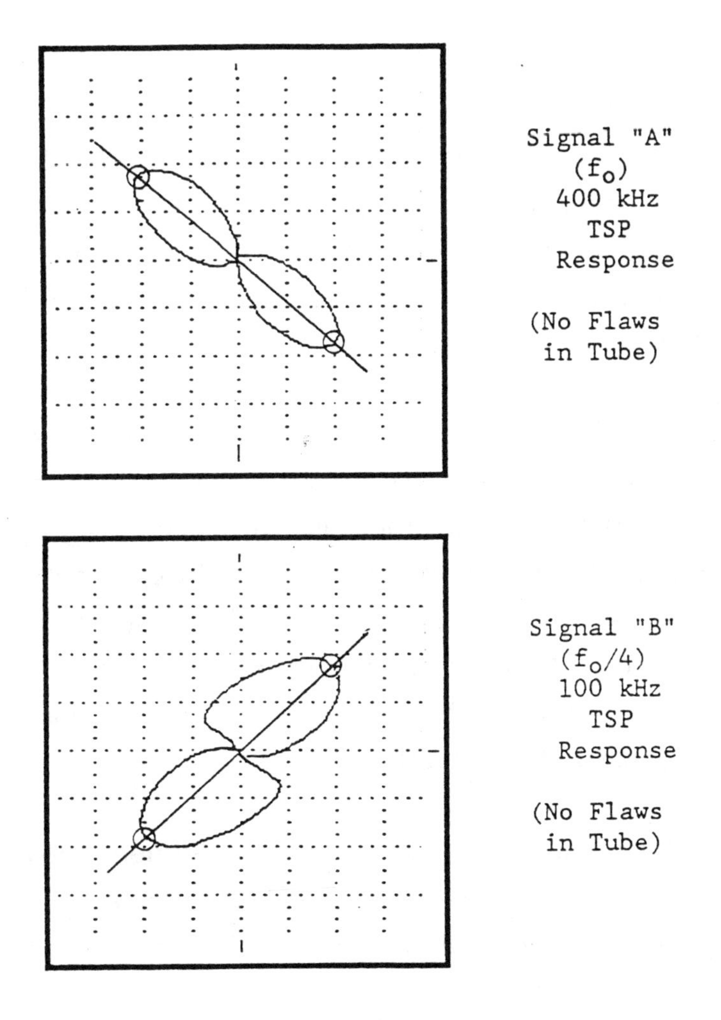

Figure 8-12
Nominal Eddy Current Response at a TSP Location at Two Different Frequencies

The flaws might be created by any one of several conditions based on either chemical attack, mechanical vibration, or the flow of water and/or steam through and around the tubes.

In Figure 8-13, let's look at the same TSP intersection that we saw in Figure 8-12, but this time there is also a 60 percent through-wall flaw (crack or pit open to the OD surface) under the TSP. The eddy currents will respond to the total field of view, which means we will see a response showing both the TSP and the flaw at the same time.

If you compare Figures 8-12 and 8-13, you will probably note that there is some signal change. The general shape of the TSP response has changed. We have to recognize that this change is due to a combination of signals from the flaw that is masked or distorted by the TSP around it.

If we could find a way to reduce the amplitude of the original TSP signal, then maybe we could see the flaw signal better. That is what mixing does for us. Electronically, we can take signals "A" and "B" and combine them to **make a new data channel,** as shown in Figure 8-14. If we measured the TSP phase angle response in these two data channels, we would see that they are approximately 90° "out of phase" with each other. This is considered a good relationship to mix or suppress non-flaw responses.

The more advanced eddy current equipment may incorporate mixers as well as many other special features to allow high precision data collection and analysis.

In nuclear power plant applications, all of the data collected from the steam generator tubes must be analyzed by multiple teams of analysts to help assure that all signals of interest are properly detected and analyzed. The training requirements, equipment capability, and probe technology must keep pace as inspection requirements become more and more sophisticated.

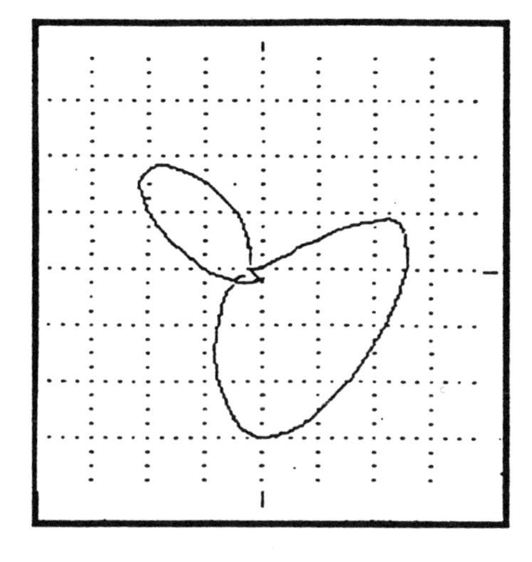

Signal "A"
(f_o)
400 kHz

Response to the TSP when a 60% FBH occurs in the tube at the same point in time.

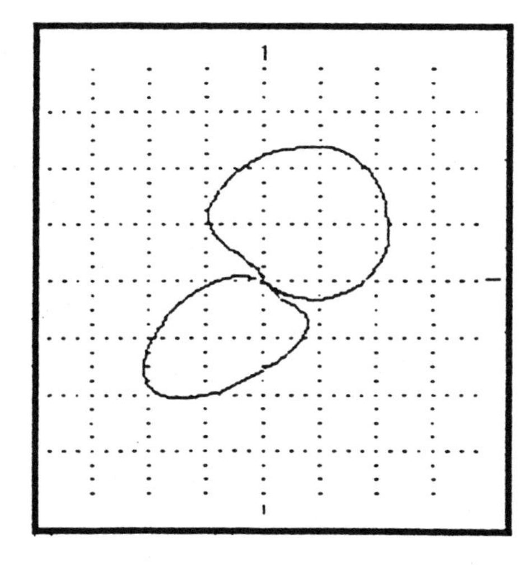

Signal "B"
($f_o/4$)
100 kHz

Response to the TSP when a 60% FBH occurs in the tube at the same point in time.

Figure 8-13
TSP Responses When a 60 Percent Through-Wall FBH Occurs at the Same Point

Modern multifrequency eddy current inspection equipment can be operated at up to four frequencies at the same time or in very rapid sequence (multiplexed). These signals can be applied to arrays of coils with up to 64 channels functioning simultaneously. The eddy current output signals are digitized, processed through various mixers and filters, and can be stored on recorders for future reference and analysis.

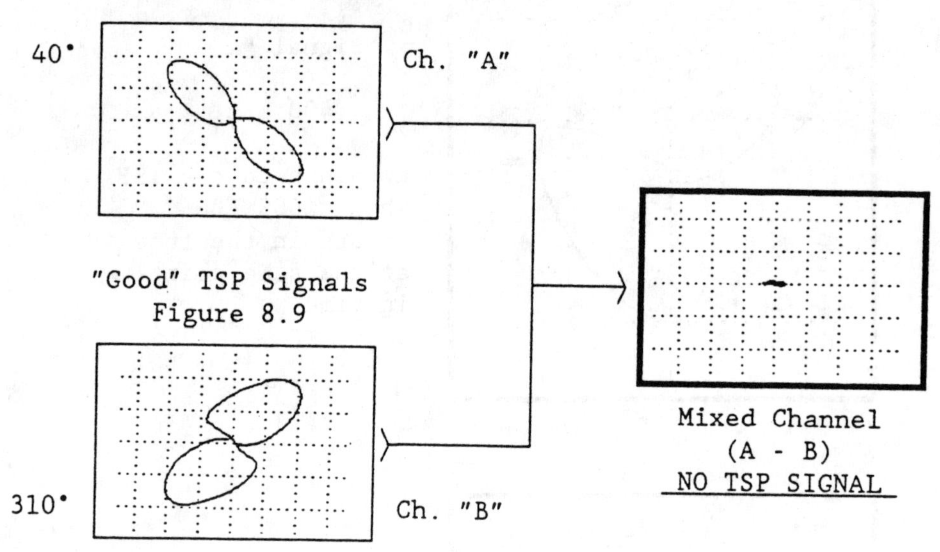

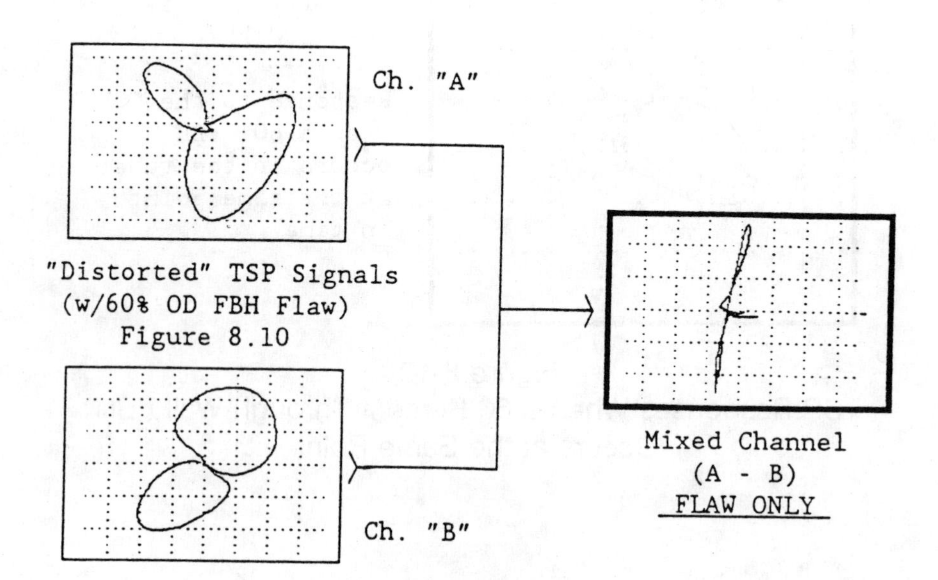

Figure 8-14
A "Mix" Channel Response to Both the
Nominal TSP and the TSP with the 60 Percent Flaw

Systems can be multitasked or dedicated. The major difference between them is what has to be done with the data after testing. They both may have fully digital, multifrequency capability. If it can be analyzed in real time (Go/No-Go) to meet the inspection requirements, then its cost may be comparable to an analog system. If full data storage and analysis capability functions are required by the inspection, then the system is going to be much more sophisticated and costlier.

In Figure 8-15, a dedicated, computer-based, multifrequency data acquisition system is schematically shown for comparison to the simpler and more portable systems used for balance-of-plant, aircraft, and general surface testing applications.

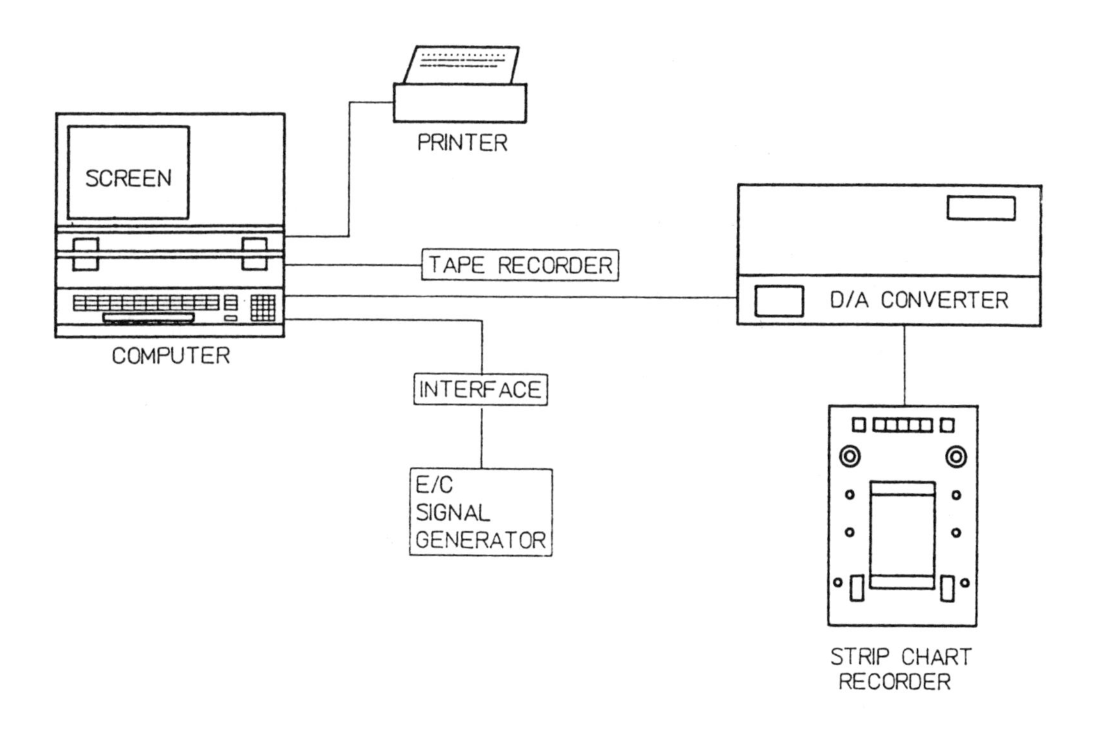

Figure 8-15
Computer-Based, Multiparameter/Multifrequency EC System

APPENDIX A

COMPARISON AND SELECTION OF NDT PROCESSES

TABLE OF CONTENTS

Page

LIST OF FIGURES

APPENDIX A

COMPARISON AND SELECTION OF NDT PROCESSES

General

This appendix summarizes the characteristics of various types of discontinuities and lists the NDT methods that may be employed to detect each type of discontinuity.

The relationship between the various NDT methods and their capabilities and limitations when applied to the detection of a specific discontinuity is shown. Such variables as type of discontinuity (inherent, process, or service), manufacturing processes (heat treating, machining, welding, grinding, or plating), and limitations (metallurgical, structural, or processing) also help in determining the sequence of testing and the ultimate selection of one test method over another.

Method Identification

Figures A-1 through A-5 illustrate five NDT methods. Each illustration shows the three elements involved in all five tests, the different methods in each test category, and tasks that may be accomplished with a specific method.

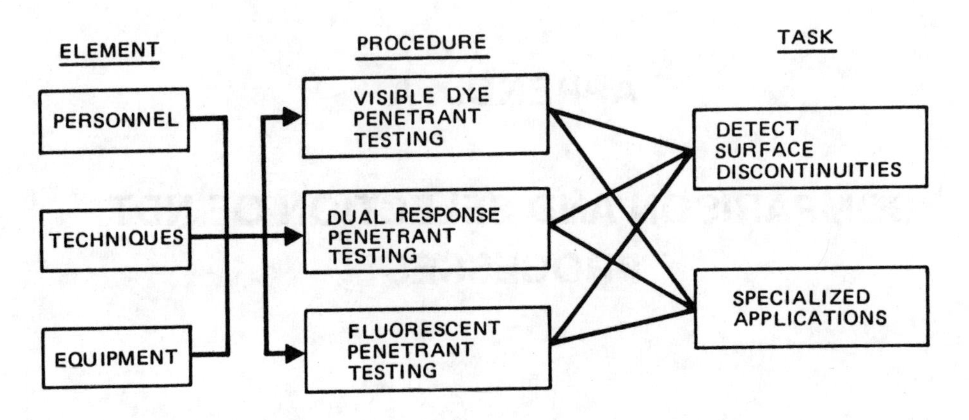

Figure A-1. Liquid Penetrant Test

NDT Discontinuity Selection

The discontinuities that are discussed in the following paragraphs are only some of the many hundreds that are associated with the various materials, processes, and products currently in use. During the selection of discontinuities for inclusion in this chapter, only those discontinuities which would not be radically changed under different conditions of design, configuration, standards, and environment were chosen.

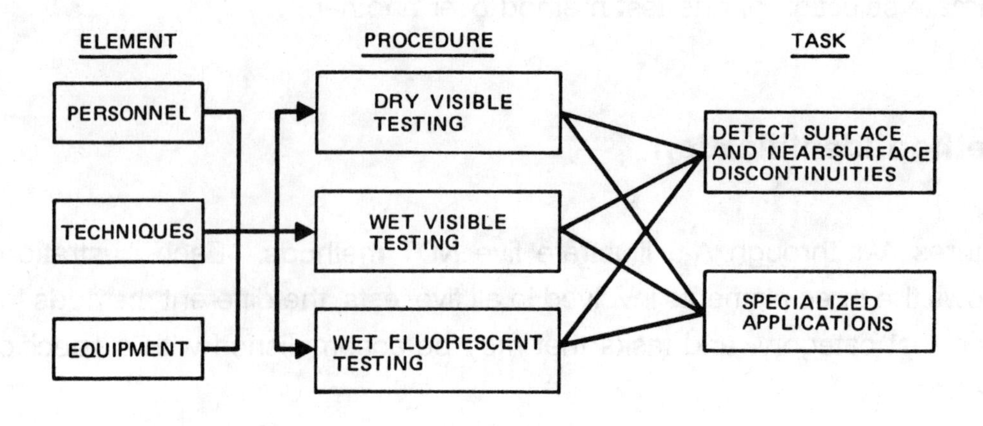

Figure A-2. Magnetic Particle Test

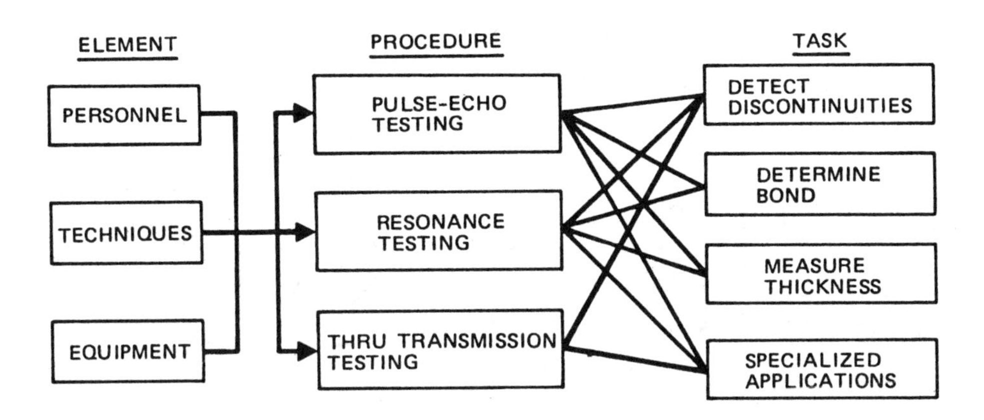

Figure A-3. Ultrasonic Test

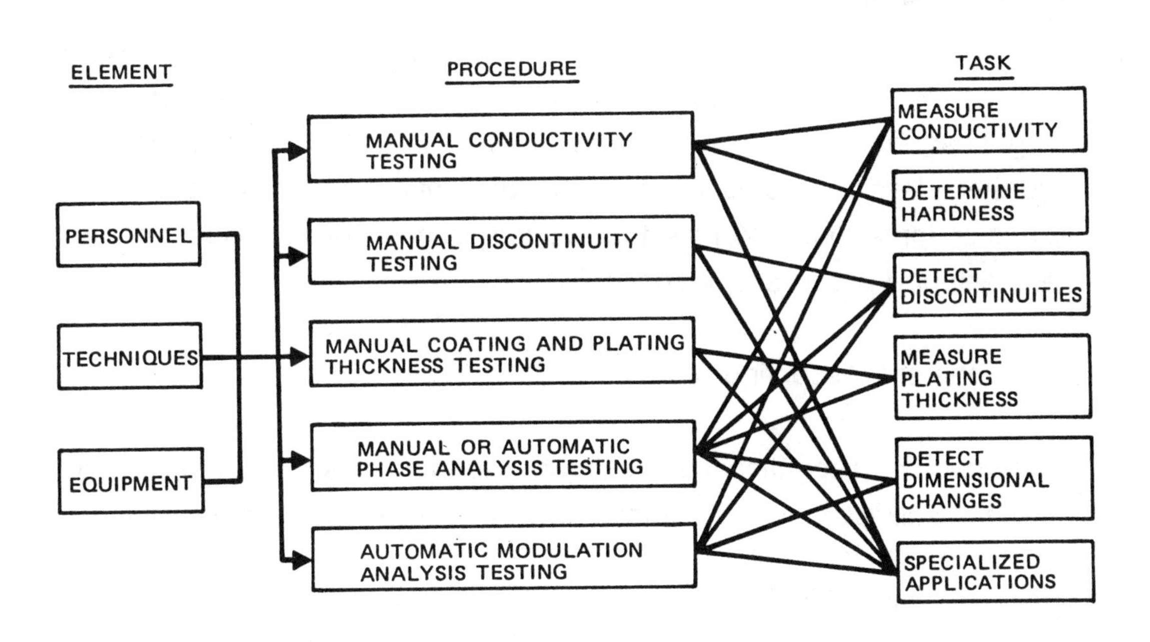

Figure A-4. Eddy Current Test

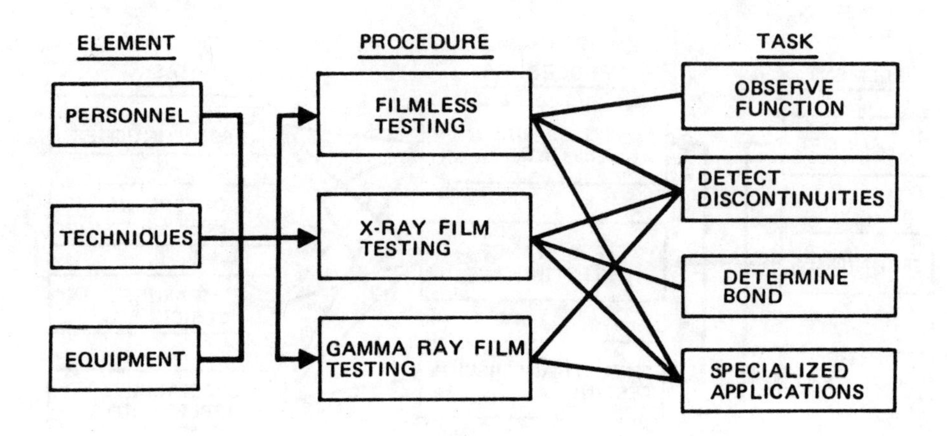

Figure A-5. Radiographic Test

Discontinuity Categories

Each of the specific discontinuities are divided into three general categories: inherent, processing, and service. Each of these categories is further classified as to whether the discontinuity is associated with ferrous or nonferrous materials, the specific material configuration, and the manufacturing processes, if applicable.

- Inherent Discontinuities

  Inherent discontinuities are those discontinuities that are related to the solidification of the molten metal. There are two types.

  - Wrought

    Inherent wrought discontinuities cover those discontinuities which are related to the melting and original solidification of the metal or ingot.

- Cast

  Inherent cast discontinuities are those discontinuities which are related to the melting, casting, and solidification of the cast article. It includes those discontinuities that would be inherent to manufacturing variables such as inadequate feeding, gating, excessively-high pouring temperature, entrapped gases, handling, and stacking.

- Processing Discontinuities

  Processing discontinuities are those discontinuities that are related to the various manufacturing processes such as machining, forming, extruding, rolling, welding, heat treating, and plating.

- Service Discontinuities

  Service discontinuities cover those discontinuities that are related to the various service conditions such as stress corrosion, fatigue, and wear.

Discontinuity Characteristics and Metallurgical Analysis

"Discontinuity characteristics," as used in this chapter, encompasses an analysis of specific discontinuities and references actual photos that illustrate examples of the discontinuity. The discussions cover the following.

- Origin and location of discontinuity (surface, near surface, or subsurface)

- Orientation (parallel or normal to the grain)

- Shape (flat, irregularly-shaped, or spiral)

- Photo (micrograph and/or typical overall view of the discontinuity)

- Metallurgical analysis (how the discontinuity is produced and at what stage of manufacture)

NDT Methods Application and Limitations

- General

  The technological accomplishments in the field of nondestructive testing have brought test reliability and reproductibility to a point where the design engineer may now selectively zone the specific article. Zoning is based upon the structural application of the end product and takes into consideration the environment as well as the loading characteristics of the article. Such an evaluation in no way reduces the end reliability of the product, but evaluation does reduce needless rejection of material that otherwise would have been acceptable. Keep in mind that the design engineer must design the most economical component(s), both in terms of cost and use of resources, that will meet the requirements of the application.

  Just as the structural application within the article varies, the allowable discontinuity size will vary, depending on the configuration and method of manufacture. For example, a die forging that has large masses of material and an extremely thin web section would not require the same level of acceptance over the entire forging. The forging can be zoned for rigid control of areas where the structural loads are higher, and less rigid for areas where the structural loads permit larger discontinuities.

The nondestructive testing specialist must also select the method which will satisfy the design objective of the specific article and not assume that all NDT methods can produce the same reliability for the same type of discontinuity.

- Selection of the NDT Method

  In selecting the NDT method for the evaluation of a specific discontinuity, keep in mind that NDT methods may supplement each other and that several NDT methods may be capable of performing the same task. The selection of one method over another is based upon such variables as those listed below.

  - Type and origin of discontinuity
  - Material manufacturing processes
  - Accessibility of article
  - Level of acceptability desired
  - Equipment available
  - Cost

  A planned analysis of the task must be made for each article requiring NDT testing.

  The NDT methods listed for each discontinuity in the following paragraphs are in order of preference for that particular discontinuity. However, when reviewing the discussions, it should be kept in mind that new techniques in the NDT field may alter the order of test preference. Literature is available from several resources that addresses many of these specialized NDT methods and techniques.

- Limitations

  The limitations applicable to the various NDT methods will vary with the applicable standard, the material, and the service environment. Limitations not only affect the NDT method but, in many cases, they also affect the structural reliability of the test article. For these reasons, limitations that are listed for one discontinuity may also be applicable to other discontinuities under slightly different conditions of material or environment. In addition, the many combinations of environment, location, materials, and test capability do not permit mentioning all limitations that may be associated with the problems of locating a specific discontinuity. The intent of this chapter is fulfilled if you are made aware of the many factors that influence the selection of a valid NDT method.

Burst

- Category - Processing

- Material - Ferrous and Nonferrous Wrought Material

- Discontinuity Characteristics

  Surface or internal. Straight or irregular cavities varying in size from wide open to very tight. Usually parallel with the grain. Found in wrought material that required forging, rolling, or extruding (Figure A-6).

A FORGING EXTERNAL BURST

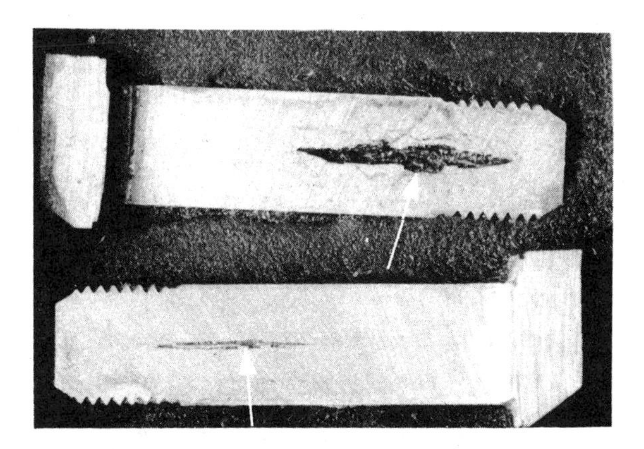

B BOLT INTERNAL BURST

C ROLLED BAR INTERNAL BURST

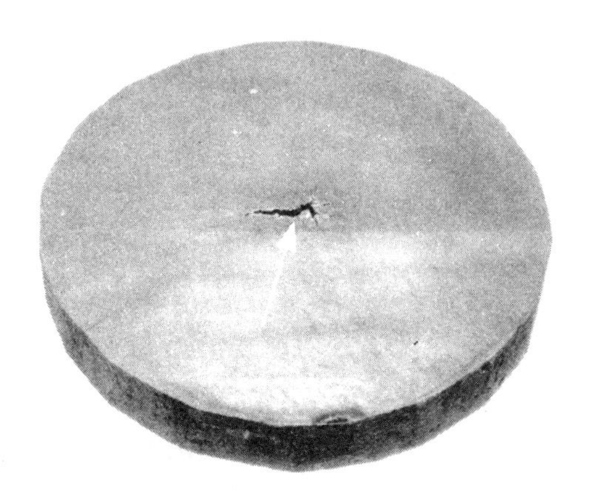

D FORGED BAR INTERNAL BURST

Figure A-6. Burst Discontinuities

- Metallurgical Analysis

  - Forging bursts are surface or internal ruptures caused by processing at too low a temperature, excessive working, or metal movement during the forging, rolling, or extruding operation.

  - A burst does not have a spongy appearance and is therefore distinguishable from a pipe, even when it occurs at the center.

  - Bursts are often large and are very seldom healed during subsequent working.

- NDT Methods Application and Limitations

  - Ultrasonic Testing Method

    - Normally used for the detection of internal bursts.

    - Bursts are definite breaks in the material and resemble a crack, producing a very sharp reflection on the scope.

    - Ultrasonic testing is capable of detecting varying degrees of burst, a condition not detectable by other NDT methods.

    - Nicks, gouges, raised areas, tool tears, foreign material, or gas bubbles on the article may produce adverse ultrasonic test results.

- Eddy Current Testing Method

    - Not normally used. Testing is restricted to wire, rod, and other articles under 0.250 inch (6.35 mm) in diameter.

- Magnetic Particle Testing Method

    - Usually used on wrought ferromagnetic material in which the burst is open to the surface or has been exposed to the surface.

    - Results are limited to surface and near surface evaluation.

- Liquid Penetrant Testing Method

    - Not normally used. When fluorescent penetrant is to be applied to an article previously dye penetrant tested, all traces of dye penetrant should first be removed by prolonged cleaning in applicable solvent.

- Radiographic Testing Method

    - Not normally used. Such variables as the direction of the burst, close interfaces, wrought material, discontinuity size, and material thickness restrict the capability of radiography.

Cold Shuts

- Category - Inherent
- Material - Ferrous and Nonferrous Cast Material
- Discontinuity Characteristics

  Surface and subsurface. Generally appear on the cast surface as smooth indentations which resemble a forging lap (Figure A-7).

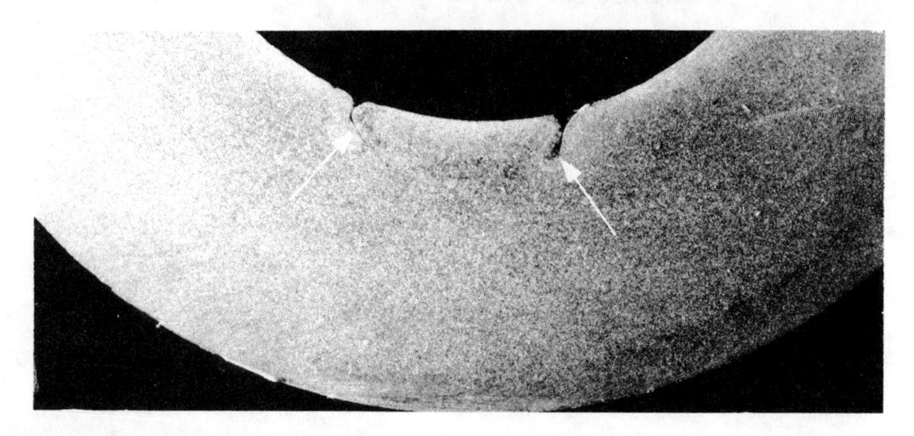

A SURFACE COLD SHUT

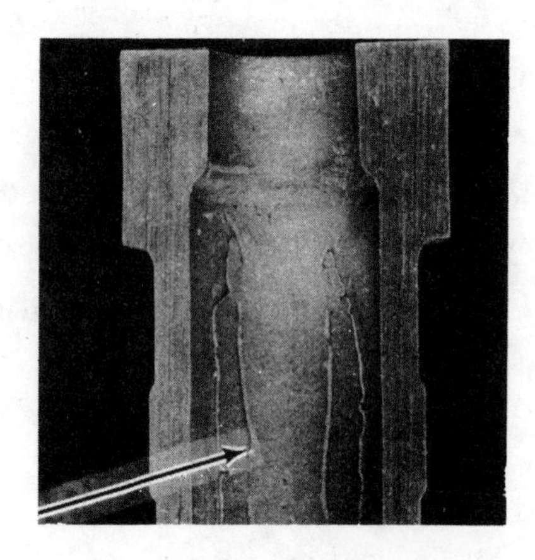

B INTERNAL COLD SHUT

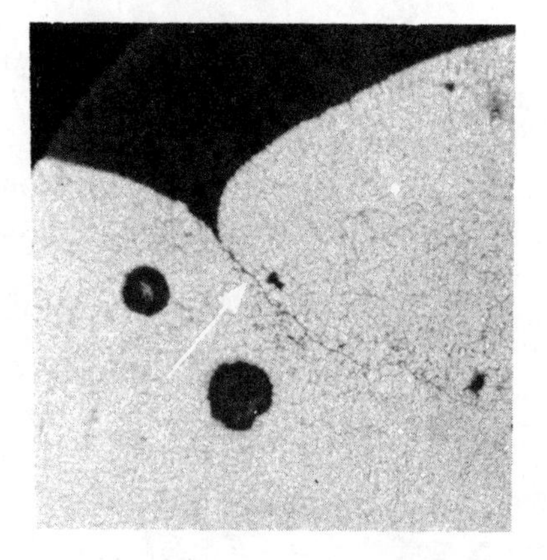

C SURFACE COLD SHUT MICROGRAPH

Figure A-7. Cold Shut Discontinuities

- Metallurgical Analysis

  Cold shuts are produced during casting of molten metal and may be caused by splashing, surging, interrupted pouring, or the meeting of two streams of metal coming from different directions. Cold shuts are also caused by the solidification of one surface before other metal flows over it, the presence of interposing surface films on cold, sluggish metal, or any factor that prevents fusion where two surfaces meet. Cold shuts are more prevalent in castings formed in a mold having several sprues or gates.

- NDT Methods Application and Limitations

  - Liquid Penetrant Testing Method

    - Normally used to evaluate surface cold shuts in both ferrous and nonferrous materials.

    - Indications appear as a smooth, regular, continuous, or intermittent line.

    - Liquid penetrants used to test nickel-based alloys, certain stainless steels, and titanium should not exceed one percent sulfur or chlorine.

    - Certain castings may have surfaces that are blind and from which removal of excess penetrant may be difficult.

    - The geometric configuration (recesses, orifices, and flanges) of a casting may permit buildup of wet developer, thereby masking any detection of a discontinuity.

- Magnetic Particle Testing Method

  - Normally used for the evaluation of ferromagnetic materials.

  - The metallurgical nature of some corrosion-resistant steel is such that, in some cases, magnetic particle testing indications are obtained which do not result from a crack or other harmful discontinuities. These indications arise from a duplex structure within the material, wherein one portion exhibits strong magnetic retentivity and the other does not.

- Radiographic Testing Method

  - Cold shuts are normally detectable by radiography while testing for other casting discontinuities.

  - Cold shuts appear as a distinct, dark line or band of variable length and width and a definite, smooth outline.

  - The casting configuration may have inaccessible areas that can only be tested by radiography.

- Ultrasonic Testing Method

  - Not recommended. As a general rule, cast structure and article configuration do not lend themselves to ultrasonic testing.

- Eddy Current Testing Method

    - Article configuration and inherent material variables require the use of specialized probes.

Fillet Cracks (Bolts)

- Category - Service

- Material - Ferrous and Nonferrous Wrought Material

- Discontinuity Characteristics

  Surface. Located at the junction of the fillet with the shank of the bolt and progressing inward (Figure A-8).

- Metallurgical Analysis

  Fillet cracks occur where a marked change in diameter occurs, such as at the head-to-shank junction where stress risers are created. During the service life of a bolt, repeated loading takes place whereby the tensile load fluctuates in magnitude due to the operation of the mechanism. These tensile loads can cause fatigue failure starting at the point where the stress risers occur. Fatigue failure, which is surface phenomenon, starts at the surface and propagates inward.

- NDT Methods Application and Limitations

  - Ultrasonic Testing Method

    - Used extensively for service-associated discontinuities of this type.

- A wide selection of transducers and equipment enable on-the-spot evaluation for fillet cracks.

- Since fillet cracks are a definite break in the material, the scope pattern will be a very sharp reflection. (Propagation can be monitored by using ultrasonics.)

- Ultrasonic equipment has extreme sensitivity, and established standards should be used to give reproducible and reliable results.

A FILLET FATIGUE FAILURE

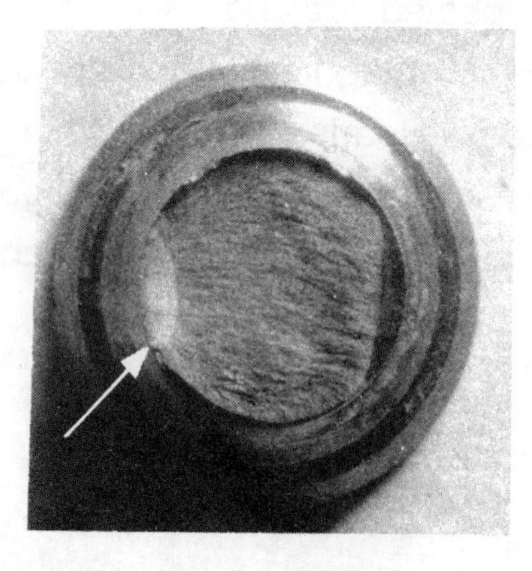

B FRACTURE AREA OF (A) SHOWING TANGENCY POINT OF FAILURE

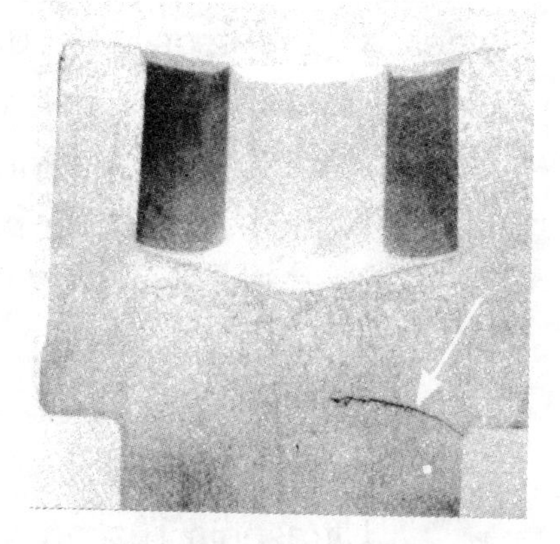

C CROSS-SECTIONAL AREA OF FATIGUE CRACK IN FILLET SHOWING TANGENCY POINT IN RADIUS

Figure A-8. Fillet Crack Discontinuity

- Liquid Penetrant Testing Method

  - Normally used during inservice overhaul or troubleshooting.

  - May be used for both ferromagnetic and nonferromagnetic bolts, although usually confined to the nonferromagnetic.

  - Fillet cracks appear as sharp, clear indications.

  - Structural damage may result from exposure of high-strength steels to paint strippers, alkaline coating removers, deoxidizer solutions, etc.

  - Entrapment of penetrant under fasteners, in holes, under splices, and in similar areas may cause corrosion due to the penetrant's affinity for moisture.

- Magnetic Particle Testing Method

  - Only used on ferromagnetic bolts.

  - Fillet cracks appear as sharp, clear indications with a heavy buildup.

  - Sharp fillet areas may produce nonrelevant magnetic indications.

  - 16.6 pH steel is only slightly magnetic in the annealed condition; however, it becomes strongly magnetic after heat treatment and can then be magnetic particle tested.

- Eddy Current Testing Method

  - Not normally used for detection of fillet cracks. Other NDT methods are more compatible to the detection of this type of discontinuity.

- Radiographic Testing Method

  - Not normally used for detection of fillet cracks. Surface discontinuities of this type would be difficult to evaluate due to the size of the crack in relation to the thickness of the material.

Grinding Cracks

- Category - Processing

- Material - Ferrous and Nonferrous

- Discontinuity Characteristics

  Surface. Very shallow and sharp at the root. Similar to heat-treat cracks and usually, but not always, occur in groups. Grinding cracks generally occur at right angles to the direction of grinding. They are found in highly heat-treated articles, chrome-plated, case-hardened, and ceramic materials that are subjected to grinding operations (Figure A-9).

- Metallurgical Analysis

  Grinding of hardened surfaces frequently introduces cracks. These thermal cracks are caused by local overheating of the surface being ground. The overheating is usually caused by lack of coolant or poor coolant, a dull or improperly ground wheel, too rapid feed or too heavy cut.

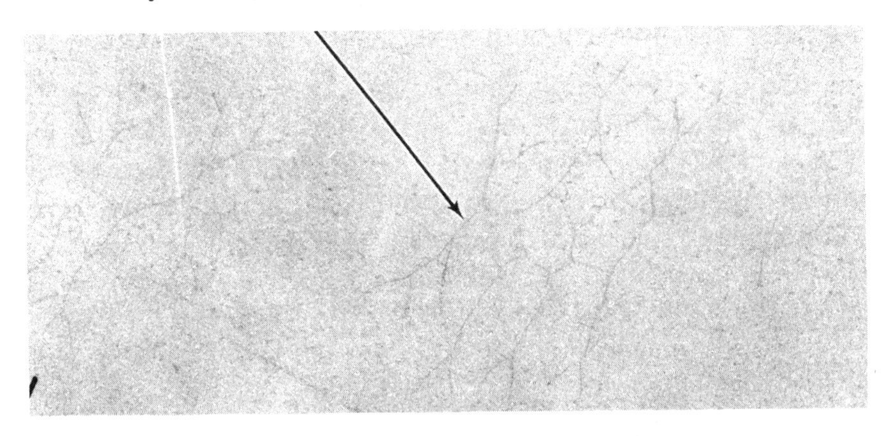

A TYPICAL CHECKED GRINDING CRACK PATTERN

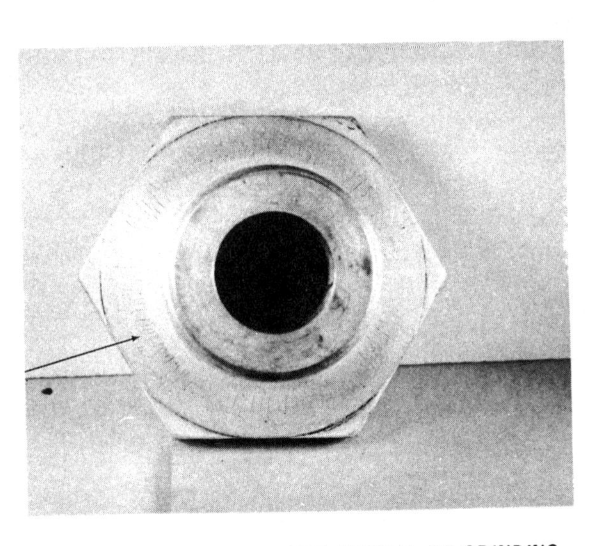

B GRINDING CRACK PATTERN NORMAL TO GRINDING

C MICROGRAPH OF GRINDING CRACK

Figure A-9. Grinding Crack Discontinuity

- NDT Methods Application and Limitations

  - Liquid Penetrant Testing Method

- Normally used on both ferrous and nonferrous materials for the detection of grinding cracks.

- Liquid penetrant indication will appear as an irregular, a checked, or a scattered pattern of fine lines.

- Grinding cracks are the most difficult discontinuities to detect and require the longest penetration time.

- Articles that have been degreased may still have solvent entrapped in the discontinuity and should be allowed sufficient time for evaporation prior to application of the penetrant.

– Magnetic Particle Testing Method

  - Restricted to ferromagnetic materials.

  - Grinding cracks generally occur at right angles to grinding direction, although in extreme cases a complete network of cracks may appear. In this case they may be parallel to the magnetic field.

  - Magnetic sensitivity decreases as the size of the grinding crack decreases.

– Eddy Current Testing Method

  - Although not normally used for detection of grinding cracks, eddy current equipment has the capability and can be developed for specific ferrous and nonferrous applications.

- Ultrasonic Testing Method

  - Not normally used for detection of grinding cracks. Other forms of NDT are more economical, faster, and better adapted to this type of discontinuity than ultrasonics.

- Radiographic Testing Method

  - Not recommended for detection of grinding cracks. Grinding cracks are too tight and too small. Other NDT methods are more suitable for detection of grinding cracks.

Convolution Cracks

- Category - Processing

- Material - Nonferrous

- Discontinuity Characteristics

  Surface. Range in size from microfractures to open fissures. Situated on the periphery of the convolutions and extend longitudinally in direction of rolling (Figure A-10).

- Metallurgical Analysis

  A rough "orange peel" effect of convolution cracks is the result of either a forming operation that stretches the material or from chemical attack, such as pickling treatment. The roughened surface contains small pits that form stress risers. Subsequent service application (vibration and flexing) may introduce stresses that act on these pits and form fatigue cracks as shown in Figure A-10.

A TYPICAL CONVOLUTION DUCTING

B CROSS-SECTION OF CRACKED CONVOLUTION

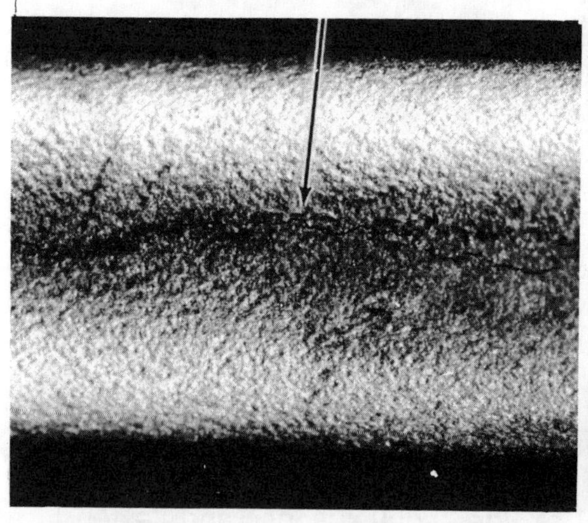

C HIGHER MAGNIFICATION OF CRACK SHOWING ORANGE PEEL

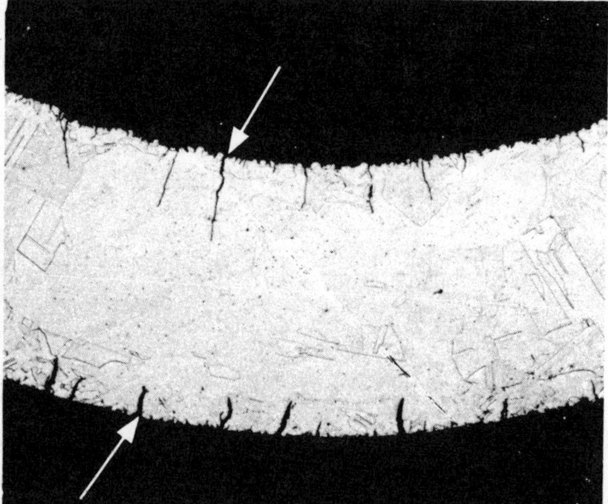

D MICROGRAPH OF CONVOLUTION WITH PARTIAL CRACKING ON SIDES

Figure A-10. Convolution Crack Discontinuities

- NDT Methods Application and Limitations

  - Radiographic Testing Method

- Used extensively for this type of failure.

- The configuration of the article and the location of the discontinuity limits detection almost exclusively to radiography.

- Orientation of convolutions to X-ray source is very critical since those discontinuities that are not normal to X-ray may not register on the film due to the small change in density.

- Liquid penetrant and magnetic particle testing may supplement, but not replace, radiographic and ultrasonic testing.

- The type of marking material (e.g., grease pencil on titanium) used to identify the area of discontinuities may affect the structure of the article.

– Ultrasonic Testing Method

  - Not normally used for the detection of convolution cracks. The configuration of the article (double-walled convolutions) and the presence of internal micro fractures are all factors that restrict the use of ultrasonics.

– Eddy Current Testing Method

  - Not normally used for the detection of convolution cracks. As in the case of ultrasonic testing, the configuration does not lend itself to this method of testing.

- Liquid Penetrant Testing Method

  - Not recommended for the detection of convolution cracks. Although the discontinuities are surface, they are internal and are superimposed over an exterior shell, which creates a serious problem of entrapment.

- Magnetic Particle Testing Method

  - Not applicable. Material is nonferrous.

Heat-Affected Zone Cracking

- Category - Processing (Weldments)

- Material - Ferrous and Nonferrous

- Discontinuity Characteristics

  Surface. Often quite deep and very tight. Usually run parallel with the weld in the heat-affected zone of the weldment (Figure A-11).

- Metallurgical Analysis

  Hot cracking or heat-affected zones of weldments increases in severity with increasing carbon content. Steels that contain more than 0.30 percent carbon are prone to this type of failure and require preheating prior to welding.

A MICROGRAPH OF WELD AND HEAT-AFFECTED ZONE SHOWING CRACK NOTE COLD LAP WHICH MASKS THE ENTRANCE TO THE CRACK

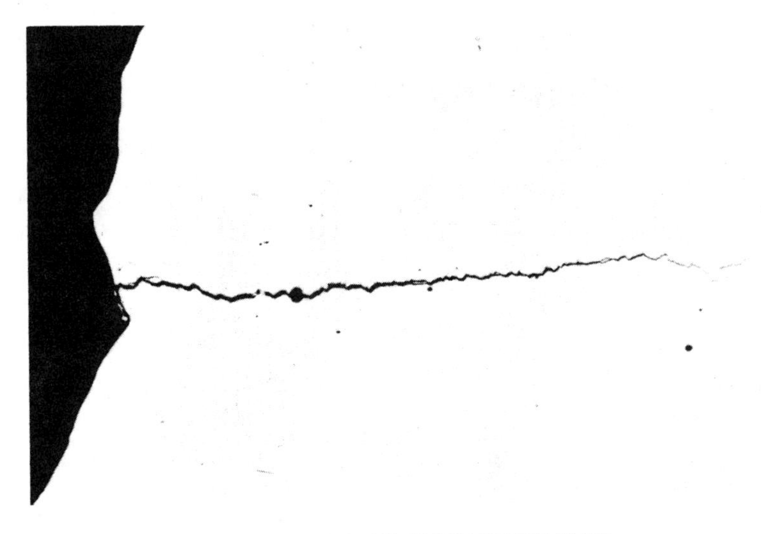

B MICROGRAPH OF CRACK SHOWN IN (A)

Figure A-11. Heat-Affected Zone Cracking Discontinuity

- NDT Methods Application and Limitations

  - Magnetic Particle Testing Method

    - Normally used for ferromagnetic weldments.

    - Prod burns are very detrimental, especially on highly-heat-treated articles. Burns may contribute to structural failure of article.

    - Demagnetization of highly-heat-treated articles can be very difficult due to metallurgical structure.

  - Liquid Penetrant Testing Method

    - Normally used for nonferrous weldments.

    - Material that has had its surface obliterated, blurred, or blended due to manufacturing processes should not be penetrant tested until the smeared surface has been removed.

    - Liquid penetrant testing after the application of certain types of chemical film coatings may be invalid due to the covering or filling of the discontinuities.

  - Radiographic Testing Method

    - Not normally used for the detection of heat-affected-zone cracking. Discontinuity orientation and surface origin make other NDT methods more suitable.

- Ultrasonic Testing Method

    - Used where specialized applications have been developed.

    - Rigid standards and procedures are required to develop valid tests.

    - The configuration of the surface roughness (i.e., sharp versus rounded root radii and the slope condition) is a major factor in deflecting the sound beam.

- Eddy Current Testing Method

    - Although not normally used for the detection of heat-affected-zone cracking, eddy current testing equipment has the capability of detecting ferrous and nonferrous surface discontinuities.

Heat-Treat Cracks

- Category - Processing

- Material - Ferrous and Nonferrous Wrought and Cast Material

- Discontinuity Characteristics

  Surface. Usually deep and forked. Seldom follow a definite pattern and can be in any direction on the part. Originate in areas with rapid change of material thickness, sharp machining marks, fillets, nicks, and discontinuities that have been exposed to the surface of the material (Figure A-12).

A FILLET AND MATERIAL THICKNESS CRACKS (TOP CENTER)
RELIEF RADIUS CRACKING (LOWER LEFT)

B HEAT-TREAT CRACK DUE TO SHARP MACHINING MARKS

Figure A-12. Heat-Treat Crack Discontinuities

- Metallurgical Analysis

During the heating and cooling process, localized stresses may be set up by unequal heating or cooling, restricted movement of the article or unequal cross-sectional thickness. These stresses may exceed the tensile strength of the material, causing it to rupture. Where built-in stress risers occur (keyways or grooves), additional cracks may develop.

- NDT Methods Application and Limitations

  - Magnetic Particle Testing Method

    - For ferromagnetic materials, heat-treat cracks are normally detected by magnetic particle testing.

    - Indications normally appear as straight, forked, or curved indications.

    - Likely points of origin are areas that would develop stress risers, such as keyways, fillets, or areas with rapid changes in material thickness.

    - Metallurgical structure of age-hardenable and heat-treatable stainless steels may produce nonrelevant indications.

  - Liquid Penetrant Testing Method

    - Liquid penetrant testing is the recommended method for nonferrous materials.

    - Likely points of origin for heat-treat cracks are the same as those listed for magnetic particle testing.

        - Materials or articles that will eventually be used in LOX systems must be tested with LOX-compatible penetrants.

    - Eddy Current Testing Method

        - Although not normally used for the detection of heat-treat cracks, eddy current testing equipment has the capability of detecting ferrous and nonferrous surface discontinuities.

    - Ultrasonic Testing Method

        - Not normally used for detection of heat-treat cracks. If used, the scope pattern will show a definite indication of a discontinuity. Recommended wave mode would be surface.

    - Radiographic Testing Method

        - Not normally used for detection of heat-treat cracks. Surface discontinuities are more easily detected by other NDT methods designed for surface application.

Surface Shrink Cracks

- Category - Processing (Welding)

- Material - Ferrous and Nonferrous

- Discontinuity Characteristics

Surface. Situated on the face of the weld, fusion zone and base metal. Range in size from very small, tight and shallow to open and deep. Cracks may run parallel or transverse to the direction of welding (Figure A-13).

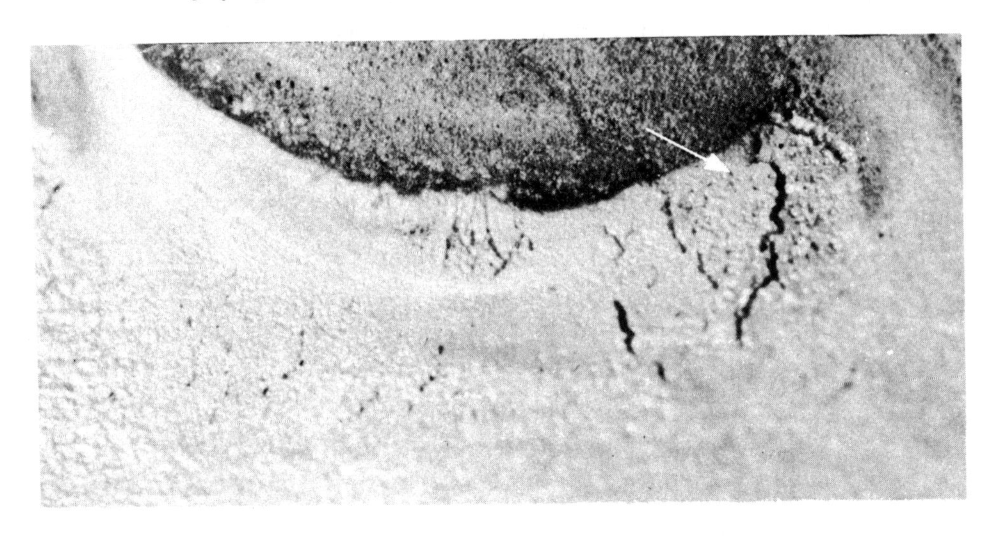

A TRANSVERSE CRACKS IN HEAT-AFFECTED ZONE

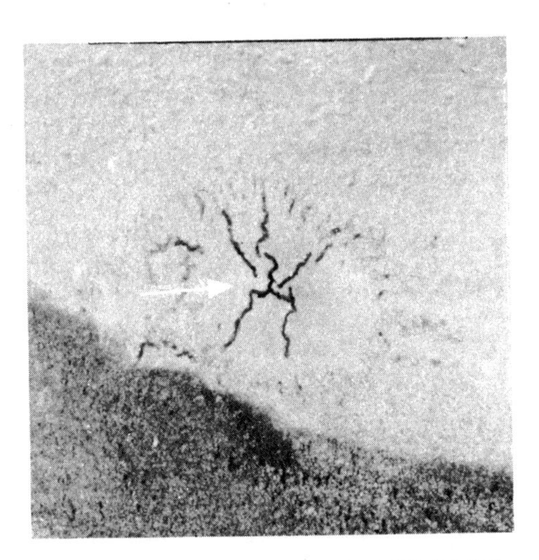

B TYPICAL STAR-SHAPED CRATER CRACK

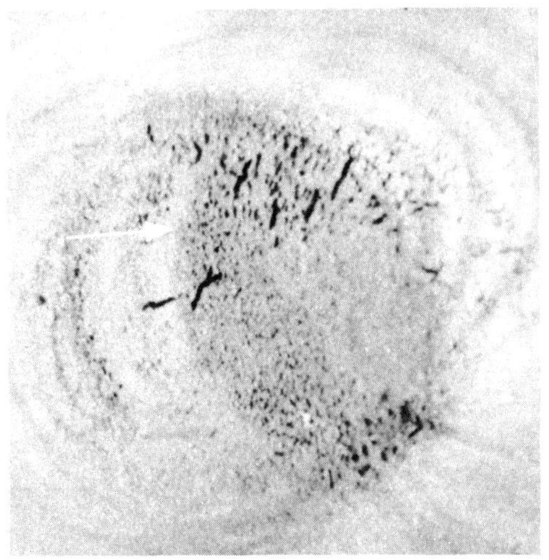

C SHRINKAGE CRACK AT WELD TERMINAL

Figure A-13. Surface Shrink Crack Discontinuities

- Metallurgical Analysis

  Surface shrink cracks are generally the result of improper heat application either in heating or welding of the article. Heating or cooling in a localized area may set up stresses that exceed the tensile strength of the material, causing the material to crack. Restriction of the movement (contraction or expansion) of the material during heating, cooling, or welding may also set up excessive stresses.

- NDT Methods Application and Limitations

  - Liquid Penetrant Testing Method

    - Surface shrink cracks in nonferrous materials are normally detected by use of liquid penetrants.

    - Liquid penetrant equipment is easily portable and can be used during in-process control for both ferrous and nonferrous weldments.

    - Assemblies that are joined by bolting, riveting, intermittent welding, or press fittings will retain the penetrant, which will seep out after developing and mask the adjoining surfaces.

    - When articles are dried in a hot air dryer or by similar means, excessive drying temperature should be avoided to prevent evaporation of penetrant.

- Magnetic Particle Testing Method

  - Ferromagnetic weldments are normally tested by magnetic particle method.

  - Surface discontinuities that are parallel to the magnetic field will not produce indications, since they do not interrupt or distort the magnetic field.

  - Areas such as grease fittings, bearing races, or other similar items that might be damaged or clogged by the bath or by the particles should be masked before testing.

- Eddy Current Testing Method

  - Ferrous and nonferrous welded sections can be inspected.

  - A probe or encircling coil could be used where article configuration permits.

- Radiographic Testing Method

  - Not normally used for the detection of surface discontinuities. During the radiographic testing of weldments for other types of discontinuities, surface indications may be detected.

- Ultrasonic Testing Method

  - Not normally used for detection of surface shrink cracks. Other forms of NDT (liquid penetrant and magnetic particle) give better results, are more economical, and are faster.

Thread Cracks

- Category - Service
- Material - Ferrous and Nonferrous Wrought Material
- Discontinuity Characteristics

  Surface. Cracks are transverse to the grain (transgranular) starting at the root of the thread (Figure A-14).

- Metallurgical Analysis

  Fatigue failures of this type are not uncommon. High cyclic stresses resulting from vibration and/or flexing act on the stress risers created by the thread roots to produce cracks. Fatigue cracks may start as fine submicroscopic discontinuities or cracks and propagate in the direction of applied stresses.

- NDT Methods Application and Limitations

  - Liquid Penetrant Testing Method

    - Fluorescent penetrant is recommended over non-fluorescent.

    - Low surface tension solvents, such as gasoline and kerosene, are not recommended cleaners.

- When applying liquid penetrant to components within an assembly or structure, the adjacent areas should be effectively masked to prevent overspraying.

A COMPLETE THREAD ROOT FAILURE

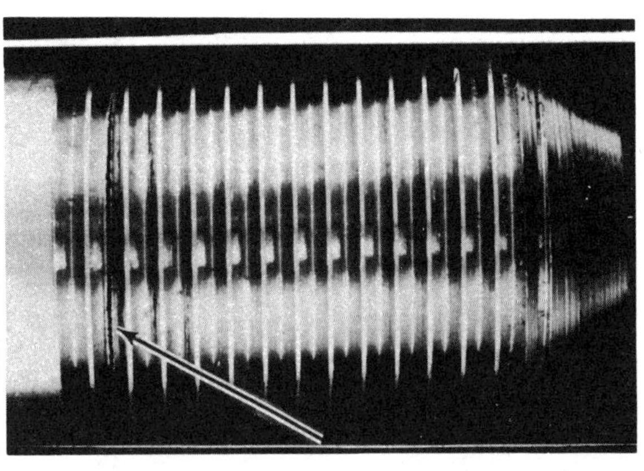

B TYPICAL THREAD ROOT FAILURE

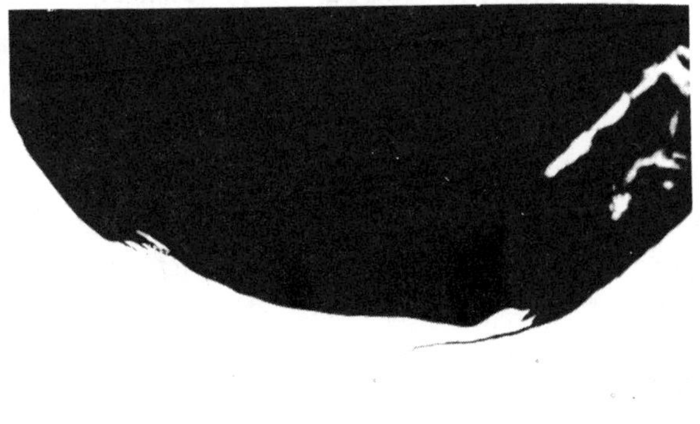

C MICROGRAPH OF (A) SHOWING CRACK AT BASE OF ROOT

D MICROGRAPH OF (B) SHOWING TRANSGRANULAR CRACK AT THREAD ROOT

Figure A-14. Thread Crack Discontinuities

- Magnetic Particle Testing Method
  - Normally used to detect cracks at the threads on ferromagnetic materials.
  - Nonrelevant magnetic indications may result from the thread configuration.
  - Cleaning titanium and 440C stainless in halogenated hydrocarbons may result in structural damage to the material.
- Ultrasonic Testing Method
  - The article configuration can be examined utilizing the cylindrical guided-wave techniques. This method requires access to the article and poses interpretation difficulties.
- Eddy Current Testing Method
  - A specialized probe to fit thread size would be required.
- Radiographic Testing Method
  - Not recommended for detecting thread cracks. Surface discontinuities are best screened by NDT method designed for the specific condition. Fatigue cracks of this type are very tight and surface-connected. Detection by radiography would be extremely difficult.

Tubing Cracks

- Category - Inherent
- Material - Nonferrous
- Discontinuity Characteristics

  Tubing cracks formed on the inner surface (ID), parallel to direction of grain flow (Figure A-15).

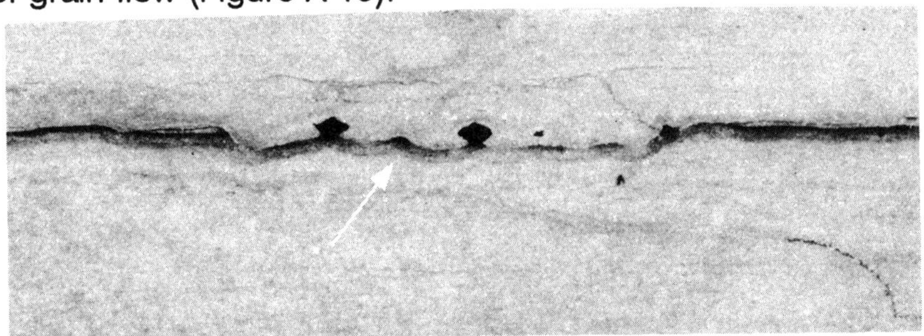

A TYPICAL CRACK ON INSIDE OF TUBING SHOWING COLD LAP

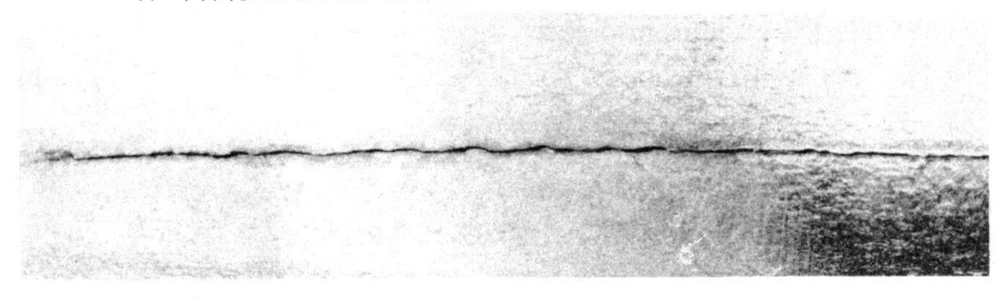

B ANOTHER PORTION OF SAME CRACK SHOWING CLEAN FRACTURE

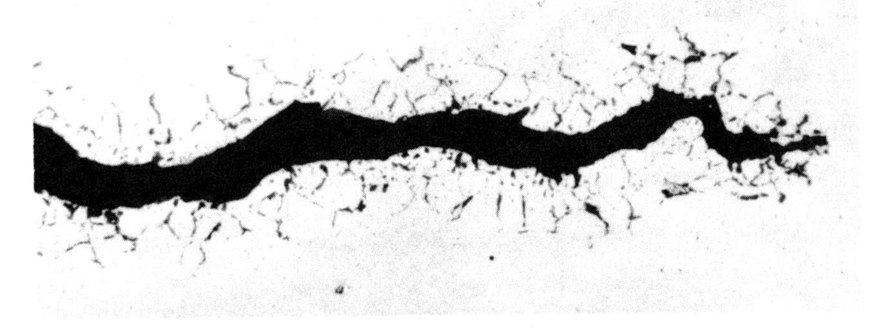

C MICROGRAPH OF (B)

Figure A-15. Tubing Crack Discontinuity

- Metallurgical Analysis

  Tubing ID cracks may be attributed to one or a combination of the following:

  - Improper cold reduction of the tube during fabrication.

  - Foreign material may have been embedded on the inner surface of the tubes causing embrittlement and cracking when the cold-worked material was heated during the annealing operation.

  - Insufficient heating rate to the annealing temperature with possible cracking occurring in the 1200°F to 1400°F (649°C to 760°C) range.

- NDT Methods Application and Limitations

  - Eddy Current Testing Method

    - Normally used for detection of this type of discontinuity.

    - Tube diameters below 1 inch (2.54 cm) and wall thicknesses less than 0.150 inch (3.8 mm) are well within equipment capability.

    - Testing of ferromagnetic material may be difficult.

  - Ultrasonic Testing Method

    - Normally used on tubing.

- A wide variety of equipment and transducers are available for screening tubing for internal discontinuities of this type.

- Ultrasonic transducers have varying temperature limitations.

- Certain ultrasonic contact couplants may have high sulfur content which will have an adverse effect on high-nickel alloys.

– Radiographic Testing Method

  - Not normally used for detecting tubing cracks. Discontinuity orientation and thickness of material govern the radiographic sensitivity. Other forms of NDT (eddy current and ultrasonics) are more economical, faster, and more reliable.

– Liquid Penetrant Testing Method

  - Not recommended for detecting tubing cracks. Internal discontinuity would be difficult to process and interpret.

– Magnetic Particle Testing Method

  - Not applicable. Material is nonferrous under normal conditions.

Hydrogen Flake

- Category - Processing

- Material - Ferrous

- Discontinuity Characteristics

  Internal fissures in a fractured surface, flakes appear as bright, silvery areas. On an etched surface they appear as short discontinuities. Sometimes known as chrome checks and hairline cracks when revealed by machining. Flakes are extremely thin and generally align parallel with the grain. They are usually found in heavy steel forging, billets, and bars (Figure A-16).

- Metallurgical Analysis

  Flakes are internal fissures attributed to stresses produced by localized transformation and decreased solubility of hydrogen during cooling after hot working. Usually found only in heavy alloy steel forgings.

- NDT Methods Application and Limitations

  - Ultrasonic Testing Method

    - Used extensively for the detection of hydrogen flake.

    - Material in the wrought condition can be screened successfully using either the immersion or the contact method. The surface condition will determine the method most suited.

A 4340 CMS HAND FORGING REJECTED FOR HYDROGEN FLAKE

B CROSS SECTION OF (A) SHOWING FLAKE CONDITION IN CENTER OF MATERIAL

Figure A-16. Hydrogen Flake Discontinuity

- On the A-scan presentation, hydrogen flake will appear as hash on the screen or as loss of back reflection.

- All foreign materials (loose scale, dirt, oil, and grease) should be removed prior to any testing. Surface irregularities such as nicks, gouges, tool marks, and scarfing may cause loss of back reflection.

– Magnetic Particle Testing Method

  - Normally used on finish machined articles.

  - Flakes appear as short discontinuities and resemble chrome checks or hairline cracks.

  - Machined surfaces with deep tool marks may obliterate the detection of the flake.

  - Where the general direction of a discontinuity is questionable, it may be necessary to magnetize in two or more directions.

– Liquid Penetrant Testing Method

  - Not normally used for detecting flakes. Discontinuities are very small and tight and would be difficult to detect by liquid penetrants.

– Eddy Current Testing Method

  - Not recommended for detecting flakes. The metallurgical structure of ferrous materials limits their adaptability to the use of eddy current testing.

- Radiographic Testing Method

  - Not recommended for detecting flakes. The size of the discontinuity and its location and orientation with respect to the material surface restricts the application of radiography.

Hydrogen Embrittlement

- Category - Processing and Service

- Material - Ferrous

- Discontinuity Characteristics

  Surface. Small, nondimensional (interface) with no orientation or direction. Found in highly-heat-treated material that has been subjected to pickling and/or plating or in material exposed to free hydrogen (Figure A-17).

- Metallurgical Analysis

  Operations such as electroplating or pickling and cleaning prior to electroplating generate hydrogen at the surface of the material. This hydrogen penetrates the surface of the material, creating immediate or delayed embrittlement and cracking.

- NDT Methods Application and Limitations

  - Magnetic Particle Testing Method

    - Magnetic indications appear as a fractured pattern.

A DETAILED CRACK PATTERN OF HYDROGEN EMBRITTLEMENT

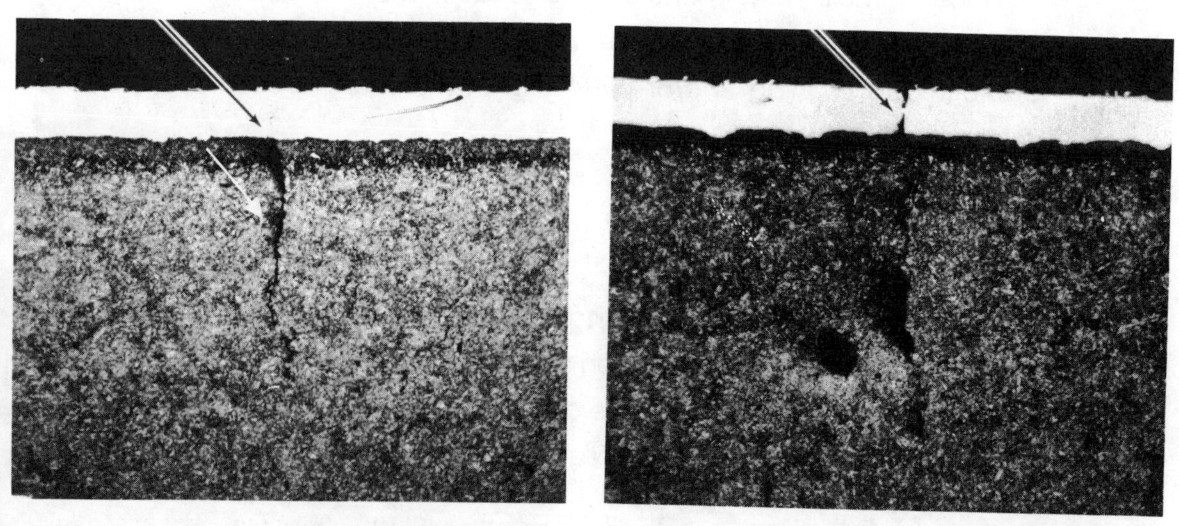

B HYDROGEN EMBRITTLEMENT UNDER CHROME PLATE C HYDROGEN EMBRITTLEMENT PROPAGATED THROUGH CHROME PLATE

Figure A-17. Hydrogen Embrittlement Discontinuity

- Hydrogen embrittlement cracks are randomly oriented and may be aligned with the magnetic field.

- Magnetic particle testing should be accomplished before and after plating.

- Care should be taken so as not to produce nonrelevant indications or cause damage to the article by overheating.

- Some alloys of corrosion-resistant steel are nonmagnetic in the annealed condition, but become magnetic with cold working.

– Liquid Penetrant Testing Method

- Not normally used for detecting hydrogen embrittlement. Discontinuities on the surface are extremely tight, small, and difficult to detect. Subsequent plating deposit may mask the discontinuity.

– Ultrasonic Testing Method

- Although ultrasonic equipment has the capability of detecting hydrogen embrittlement, this method is not normally used. Article configurations and size do not, in general, lend themselves to this method of testing. Surface wave and/or time-of-flight techniques are recommended.

- Eddy Current Testing Method

  - Not recommended for detecting hydrogen embrittlement. Many variables inherent in the specific material may produce conflicting patterns.

- Radiographic Testing Method

  - Not recommended for detecting hydrogen embrittlement. The sensitivity required to detect hydrogen embrittlement is, in most cases, in excess of radiographic capabilities.

Inclusions

- Category - Processing (Weldments)

- Material - Ferrous and Nonferrous Welded Material

- Discontinuity Characteristics

  Surface and subsurface. Inclusions may be any shape. They may be metallic or nonmetallic and may appear individually or be linearly distributed or scattered throughout the weldment (Figure A-18).

- Metallurgical Analysis

  Metallic inclusions are generally particles of metals of different density as compared to the density of the weld or base metal. Nonmetallic inclusions are oxides, sulphides, slag, or other nonmetallic foreign material entrapped in the weld or trapped between the weld metal and the base metal.

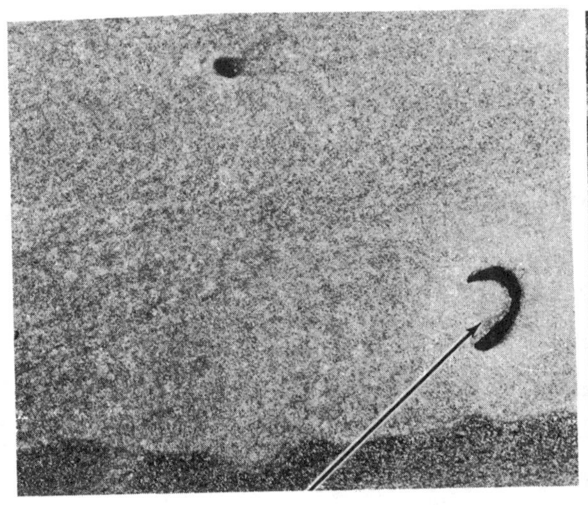

A METALLIC INCLUSIONS

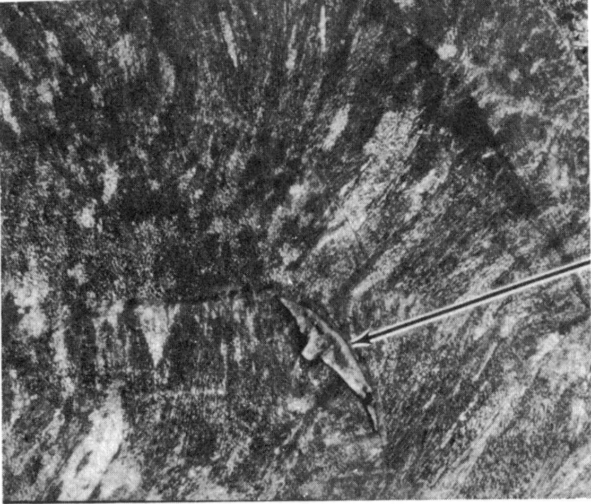

B INCLUSIONS TRAPPED IN WELD

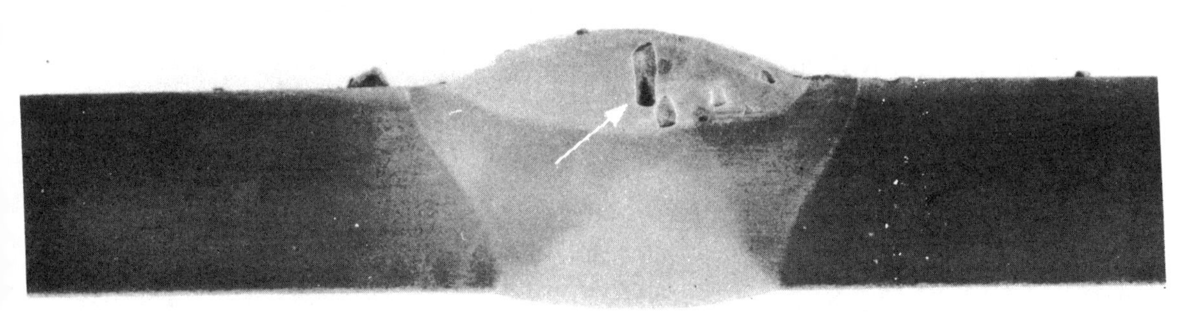

C CROSS SECTION OF WELD SHOWING INTERNAL INCLUSIONS

Figure A-18. Weldment Inclusion Discontinuities

- NDT Methods Application and Limitations
  - Radiographic Testing Method
    - This NDT method is universally used.
    - Metallic inclusions appear on the radiograph as sharply-defined, round, erratically-shaped or elongated white spots and may be isolated or in small linear or scattered groups.

- Nonmetallic inclusions will appear on the radiograph as shadows of round globules or elongated or irregularly-shaped contours occurring individually, linearly or scattered throughout the weldment. They will generally appear in the fusion zone or at the root of the weld. Less absorbent material is indicated by a greater film density and more absorbent materials by a lighter film density.

- Foreign material such as loose scales, splatter, or flux may invalidate test results.

– Eddy Current Testing Method

- Normally confined to thin-walled welded tubing.

- Established standards are required if valid results are to be obtained.

– Magnetic Particle Testing Method

- Normally not used for detecting inclusions in weldments.

- Confined to machined weldments where the discontinuities are surface or near surface.

- The indications would appear jagged, irregularly-shaped, individually, or clustered and would not be too pronounced.

- Discontinuities may go undetected when improper contact exists between the magnetic particles and the surface of the article.

– Ultrasonic Testing Method

  - Not normally used for detecting inclusions. Specific applications of design or of article configuration, however, may require ultrasonic testing.

– Liquid Penetrant Testing Method

  - Not applicable. Inclusions are normally not open fissures.

Inclusions

- Category - Processing

- Material - Ferrous and Nonferrous Wrought Material

- Discontinuity Characteristics

  Subsurface (original bar) or surface (after machining). There are two types; one is nonmetallic with long, straight lines parallel to flow lines and quite tightly adherent. They are often short and likely to occur in groups. The other type is nonplastic, appearing as a comparatively large mass not parallel to flow lines. Found in forged, extruded, and rolled material (Figure A-19).

A TYPICAL INCLUSION PATTERN ON MACHINED SURFACES

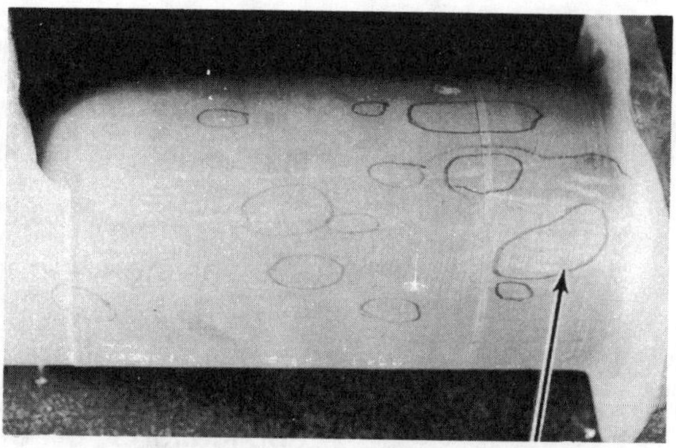

B STEEL FORGING SHOWING NUMEROUS INCLUSIONS

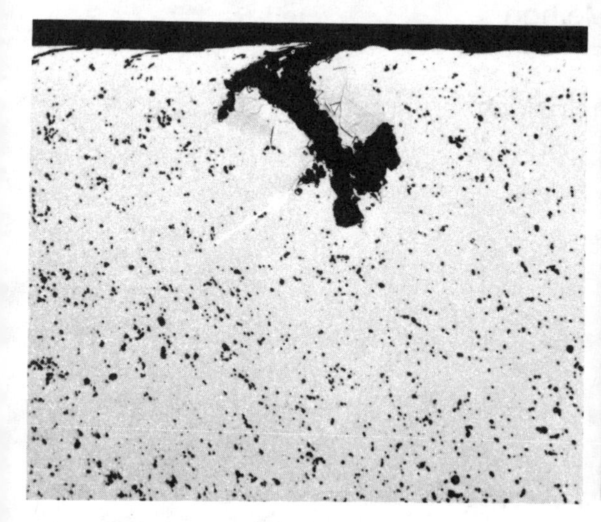

C MICROGRAPH OF TYPICAL INCLUSION

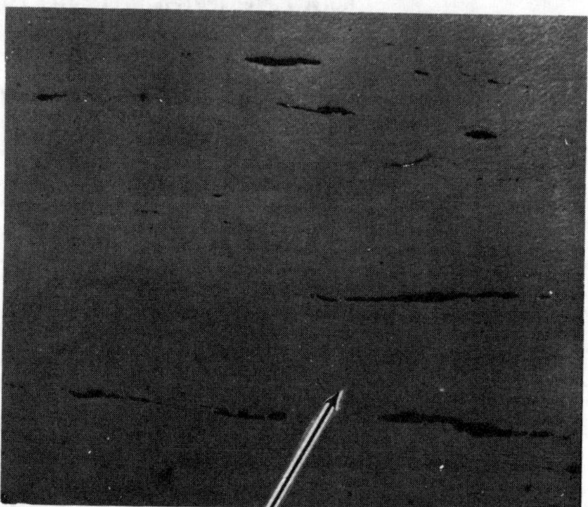

D LONGITUDINAL CROSS SECTION SHOWING ORIENTATION OF INCLUSIONS

Figure A-19. Wrought Inclusion Discontinuities

- Metallurgical Analysis

Nonmetallic inclusions (stringers) are caused by the existence of slag or oxides in the billet or ingot. Nonplastic inclusions are caused by particles remaining in the solid state during billet melting. Certain types of steels are more prone to inclusions than others.

- NDT Methods Application and Limitations
  - Ultrasonic Testing Method
    - Normally used to evaluate inclusions in wrought material.
    - Inclusions will appear as definite interfaces within the metal. Small, clustered condition or conditions on different planes cause a loss in back reflection. Numerous small, scattered conditions cause excessive "noise."
    - Inclusion orientation in relationship to ultrasonic beam is critical.
    - The direction of the ultrasonic beam should be perpendicular to the direction of the grain flow whenever possible.
  - Eddy Current Testing Method
    - Normally used for thin-walled tubing and small-diameter rods.
    - Eddy current testing of ferromagnetic materials can be difficult.
  - Magnetic Particle Testing Method
    - Normally used on machined surface.

- Inclusions will appear as a straight, intermittent, or a continuous indication. They may be individual or clustered.

- The magnetizing technique should be such that a surface or near surface inclusion can be satisfactorily detected when its axis is in any direction.

- A knowledge of the grain flow of the material is critical since inclusions will be parallel to that direction.

– Liquid Penetrant Testing Method

  - Not normally used for detecting inclusions in wrought material. Inclusions are generally not openings in the material surface.

– Radiographic Testing Method

  - Not recommended. NDT methods designed for surface testing are more suitable for detecting surface inclusions.

Lack of Penetration

- Category - Processing

- Material - Ferrous and Nonferrous Weldments

- Discontinuity Characteristics

Internal or external. Generally irregular and filamentary occurring at the root and running parallel with the weld (Figure A-20).

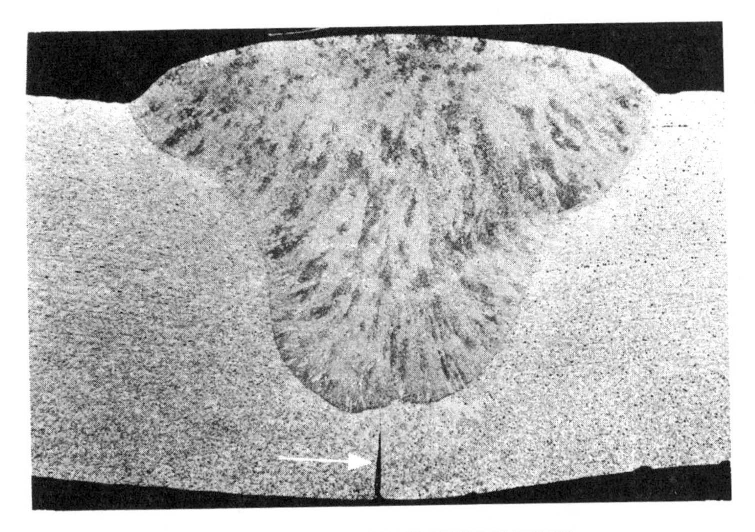

A INADEQUATE ROOT PENETRATION

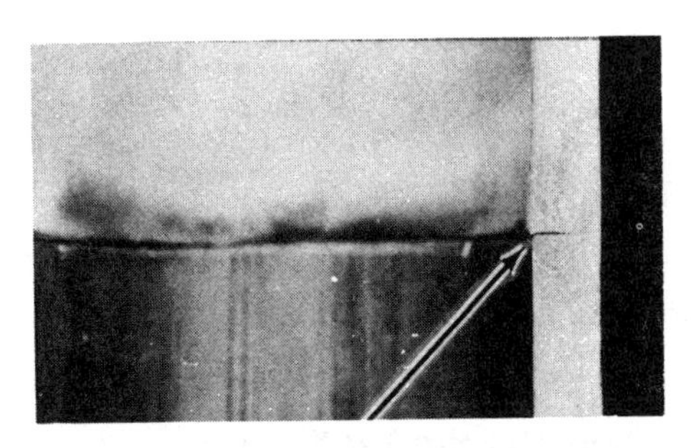

B INADEQUATE ROOT PENETRATION OF BUTT WELDED TUBE

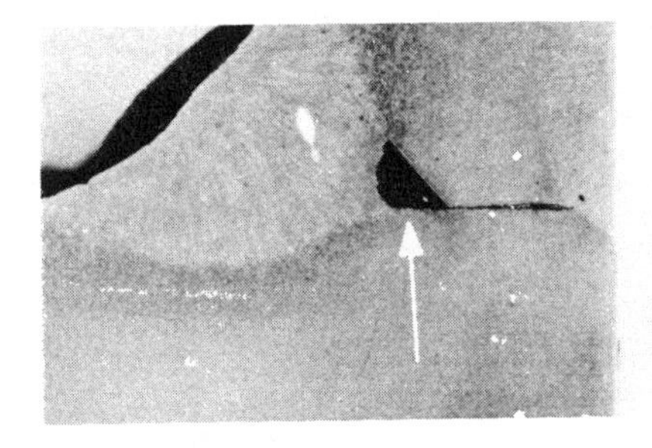

C INADEQUATE FILLET WELD PENETRATION KNOWN AS BRIDGING

Figure A-20. Lack of Penetration Discontinuities

- Metallurgical Analysis

Caused by root face of joint not reaching fusion temperature before weld metal was deposited. Also caused by fast welding rate, too large a welding rod or too cold a bead.

- NDT Methods Application and Limitations

  - Radiographic Testing Method

    - Used extensively on a wide variety of welded articles to determine the lack of penetration.

    - Lack of penetration will appear on the radiograph as an elongated, dark area of varying length and width. Lack of penetration may be continuous or intermittent and may appear in the center of the weld at the junction of multipass beads.

    - Lack of penetration orientation in relationship to the radiographic source is critical.

    - Sensitivity levels govern the capability to detect small or tight discontinuities.

  - Ultrasonic Testing Method

    - Commonly used for specific applications.

    - Weldments make ultrasonic testing difficult.

    - Lack of penetration will appear on the scope as a definite break or discontinuity resembling a crack and will give a very sharp reflection.

  - Eddy Current Testing Method

    - Normally used to determine lack of penetration in nonferrous welded pipe and tubing.

        - Eddy current testing can be used where other nonferrous articles can meet the configuration requirement of the equipment.

    - Magnetic Particle Testing Method

        - Normally used where back side of weld is visible.

        - Lack of penetration appears as an irregular indication of varying width.

    - Liquid Penetrant Testing Method

        - Normally used where backside of weld is visible.

        - Lack of penetration appears as an irregular indication of varying width.

        - Residue left by the penetrant and the developer could contaminate any rewelding operation.

Laminations

- Category - Inherent

- Material - Ferrous and Nonferrous Wrought Material

- Discontinuity Characteristics

  Surface and internal. Flat, extremely thin, generally aligned parallel to the work surface of the material. May contain a thin film of oxide between the surfaces. Found in forged, extruded and rolled material (Figure A-21).

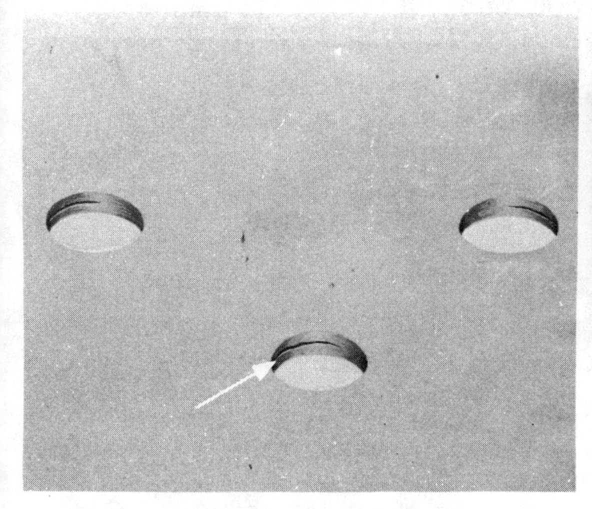

A LAMINATION IN 0.250 IN. PLATE

B LAMINATION IN 0.040 TITANIUM SHEET

C LAMINATION IN PLATE SHOWING SURFACE ORIENTATION

D LAMINATION IN 1 IN. BAR SHOWING SURFACE ORIENTATION

Figure A-21. Lamination Discontinuities

- Metallurgical Analysis

  Laminations are separations or weaknesses generally aligned parallel to the work surface of the material. They may be the result of pipe, blister, seam, inclusions, or segregations elongated and made directional by working. Laminations are flattened impurities that are extremely thin.

- NDT Methods Application and Limitations

  - Ultrasonic Testing Method

    - For heavier gauge material, the geometry and orientation of lamination (normal to the beam) makes their detection limited to ultrasonic testing.

    - Numerous wave modes may be used, depending upon the material thickness or method selected for testing. Automatic and manual contact or immersion methods are adaptable.

    - Laminations appear as a definite interface with a loss of back reflection.

    - Through-transmission and reflection techniques are applicable for very thin sections.

  - Magnetic Particle Testing Method

    - Articles fabricated from ferromagnetic materials are normally tested for lamination by magnetic particle testing methods.

- Magnetic indication will appear as a straight, intermittent indication.

- Magnetic particle testing is not capable of determining the overall size or depth of the lamination.

- Liquid Penetrant Testing Method

  - Normally used on nonferrous materials.

  - Machining, honing, lapping, or blasting may smear the surface of the material and thereby close or mask surface lamination.

  - Acid and alkalines seriously limit the effectiveness of liquid penetrant testing. Thorough cleaning of the surface is essential.

- Eddy Current Testing Method

  - Not normally used to detect laminations.

- Radiographic Testing Method

  - Not recommended for detecting laminations. Laminations have very small thickness changes in the direction of the X-ray beam thereby making radiographic detection almost impossible.

Laps and Seams

- Category - Processing

- Material - Ferrous and Nonferrous Rolled Threads

- Discontinuity Characteristics

  Surface. Wavy lines, often quite deep and sometimes very tight, appearing as hairline cracks. Found in rolled threads in the minor pitch and major diameter of the thread and in direction of rolling (Figure A-22).

- Metallurgical Analysis

  During the rolling operation, faulty or oversized dies or an overfill of material may cause material to be folded over and flattened into the surface of the thread but not fused.

- NDT Methods Application and Limitations

  - Liquid Penetrant Testing Method

    - Compatibility with both ferrous and nonferrous materials makes fluorescent liquid penetrant the first choice.

    - Liquid penetrant indications will be circumferential, slightly curved, intermittent, or continuous indications. Laps and seams may occur individually or in clusters.

A TYPICAL AREAS OF FAILURE LAPS AND SEAMS

B FAILURE OCCURRING AT ROOT OF THREAD

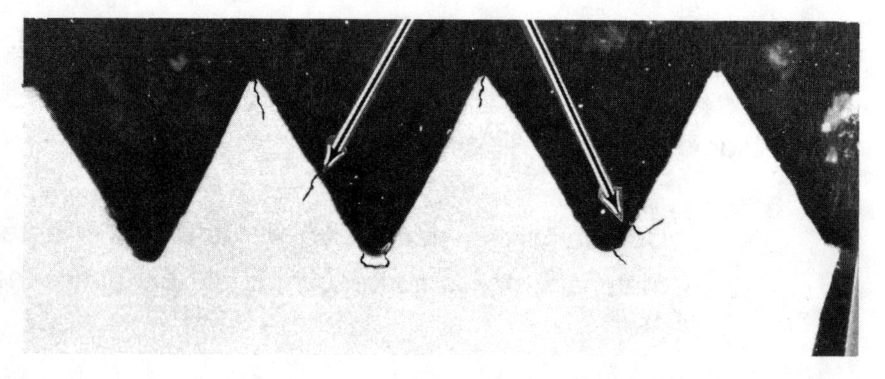

C AREAS WHERE LAPS AND SEAMS USUALLY OCCUR

Figure A-22. Lap and Seam Discontinuities in Rolled Threads

- Foreign material may not only interfere with the penetration of the penetrant into the discontinuity, but may cause an accumulation of penetrant in a nondefective area.

- Surface of threads may be smeared due to rolling operation, thereby sealing off laps and seams.

- Fluorescent and dye penetrants are not compatible. Dye penetrants tend to kill the fluorescent qualities in fluorescent penetrants.

– Magnetic Particle Testing Method

  - Magnetic particle indications of laps and seams generally appear the same as liquid penetrant indications.

  - Nonrelevant magnetic indications may result from threads.

  - Questionable magnetic particle indications can be verified by liquid penetrant testing.

– Eddy Current Testing Method

  - Probe coil design must match sample geometry.

– Ultrasonic Testing Method

  - Not recommended for detecting laps and seams. Thread configurations restrict ultrasonic capability.

- Radiographic Testing Method

    - Not recommended for detecting laps and seams. Size and orientation of discontinuities restrict the capability of radiographic testing.

Laps and Seams

- Category - Processing

- Material - Ferrous and Nonferrous Wrought Material

- Discontinuity Characteristics

    - Lap Surface. Wavy lines which are usually not very pronounced nor tightly adherent since they usually enter the surface at a small angle. Laps may have surface openings which are smeared closed. Found in wrought forgings, plate, tubing, bar, and rod (Figure A-23).

    - Seam Surface. Lengthy, often quite deep, and sometimes very tight. Usually occur in fissures parallel with the grain, and, when associated with rolled rod and tubing, they may at times be spiral.

- Metallurgical Analysis

    Seams originate from blowholes, cracks, splits, and tears introduced in earlier processing and elongated in the direction of rolling or forging. The distance between adjacent interfaces of the discontinuity is very small.

A TYPICAL FORGING LAP

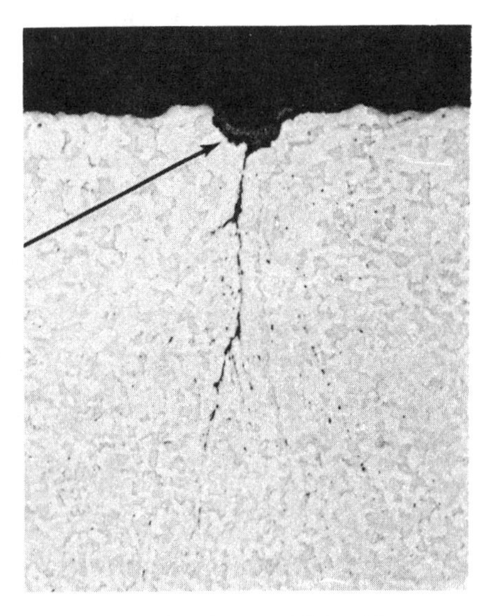

B MICROGRAPH OF A LAP

Figure A-23. Lap and Seam Discontinuities in Wrought Material

Laps are similar to seams and may result from improper rolling, forging, or sizing operations. Corners may be folded over during the processing of the material or an overfill may exist during sizing that results in the material being flattened but not fused into the surface. Laps may occur on any part of the article.

- NDT Methods Application and Limitations
  - Magnetic Particle Testing Method
    - Magnetic particle testing is recommended for ferromagnetic material.
    - Surface and near-surface laps and seams may be detected by this method.

- Laps and seams may appear as straight, spiral, or slightly curved indications. They may be individual or clustered and continuous or intermittent.

- Magnetic buildup at laps and seams is very small; therefore, a magnetizing current greater than that used for the detection of cracks is necessary.

- Correct magnetizing technique should be used when examining for forging laps since the discontinuity may lie in a plane nearly parallel to the surface.

– Liquid Penetrant Testing Method

  - Liquid penetrant testing is recommended for nonferrous material.

  - Laps and seams may be very tight and difficult to detect, especially by liquid penetrant.

  - Liquid penetrant testing of laps and seams can be improved slightly by heating the article before applying the penetrant.

– Ultrasonic Testing Method

  - Normally used to test wrought material prior to machining.

  - Surface wave and/or time-of-flight techniques permit accurate evaluation of the depth, length, and size of laps and seams.

- Ultrasonic indications of laps and seams will appear as definite interfaces within the metal.

- Eddy Current Testing Method

  - Normally used for the evaluation of laps and seams in tubing and pipe.

  - Other articles can be screened by eddy current where article configuration and size permit.

- Radiographic Testing Method

  - Not recommended for detecting laps and seams in wrought material.

Microshrinkage

- Category - Processing

- Material - Magnesium Casting

- Discontinuity Characteristics

  Internal. Small filamentary voids in the grain boundaries appear as concentrated porosity in cross section (Figure A-24).

- Metallurgical Analysis

  Shrinkage occurs while the metal is in a plastic or semimolten state. If sufficient molten metal cannot flow into different areas as it cools, the shrinkage will leave a void. The void is identified by its appearance and by the time it occurs in the plastic range. Microshrinkage is caused by the withdrawal of the low melting point constituent from the grain boundaries.

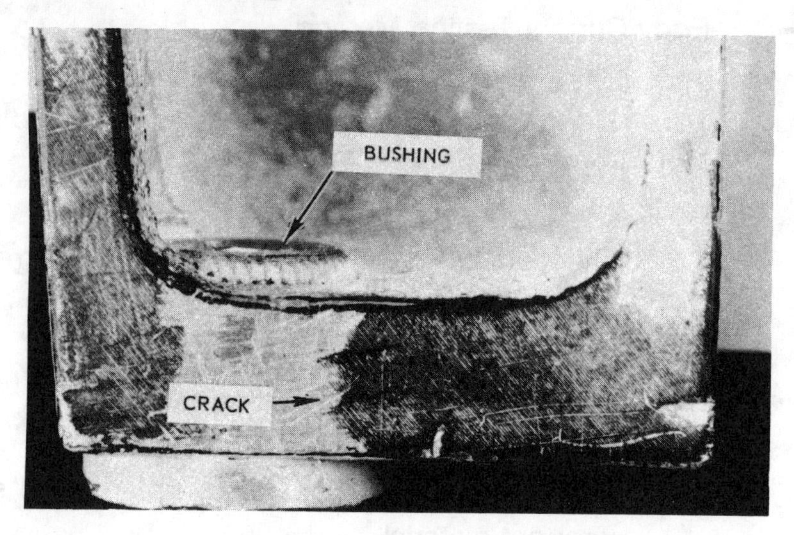

A CRACKED MAGNESIUM HOUSING

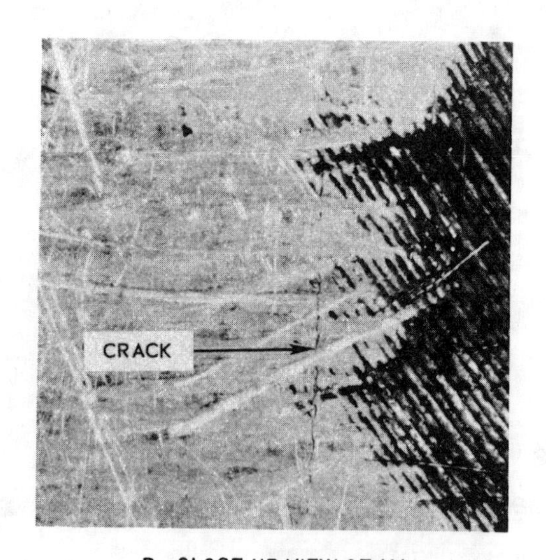

B CLOSE-UP VIEW OF (A)

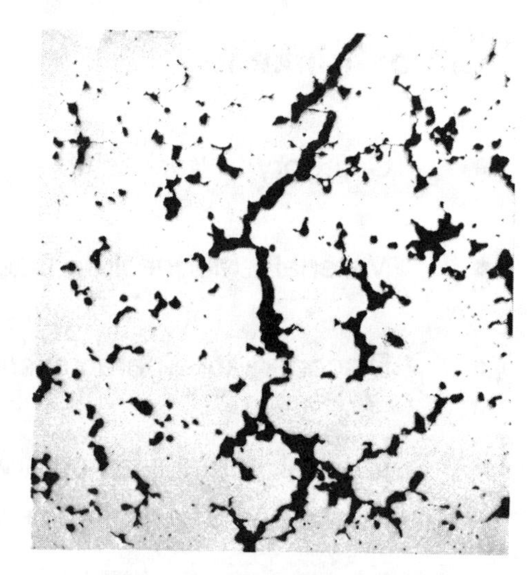

C MICROGRAPH OF CRACKED AREA

Figure A-24. Microshrinkage Discontinuity

- NDT Methods Application and Limitations

  - Radiographic Testing Method

    - Radiography is universally used to determine the acceptance level of microshrinkage.

    - Microshrinkage will appear on the radiograph as an elongated swirl resembling feathery streaks or as dark, irregular patches that are indicative of cavities in the grain boundaries.

  - Liquid Penetrant Testing Method

    - Normally used on finished machined surfaces.

    - Microshrinkage is not normally open to the surface; therefore, these conditions will be detected in machined areas.

    - The appearance of the indication depends on the plane through which the microshrinkage has been cut. The appearance varies from a continuous hairline to a massive porous indication.

    - Penetrant may act as a contaminant by saturating the microporous casting, affecting its ability to accept a surface treatment.

    - Serious structural or dimensional damage to the article can result from the improper use of acids or alkalies. They should never be used unless approval is obtained.

- Eddy Current Testing Method

  - Not recommended for detecting microshrinkage. Article configuration and type of discontinuity do not lend themselves to eddy current testing.

- Ultrasonic Testing Method

  - Not recommended for detecting microshrinkage. Cast structure and article configuration are restricting factors.

- Magnetic Particle Testing Method

  - Not applicable. Material is nonferrous.

Gas Porosity

- Category - Processing

- Material - Ferrous and Nonferrous Weldments

- Discontinuity Characteristics

  Surface or subsurface. Rounded or elongated, teardrop-shaped, with or without a sharp discontinuity at the point. Scattered uniformly throughout the weld or isolated in small groups. May also be concentrated at the root or toe (Figure A-25).

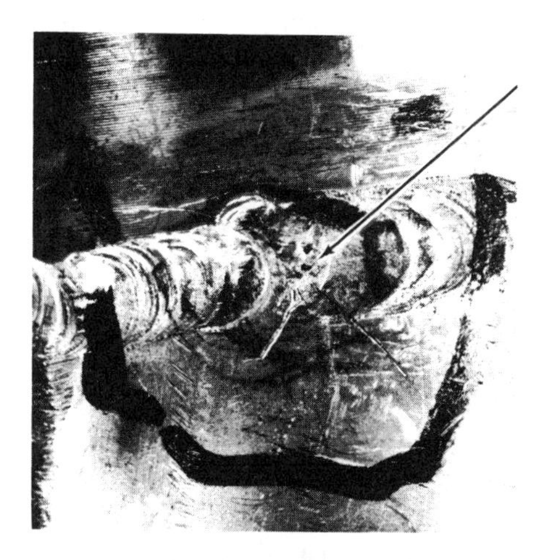

A TYPICAL SURFACE POROSITY

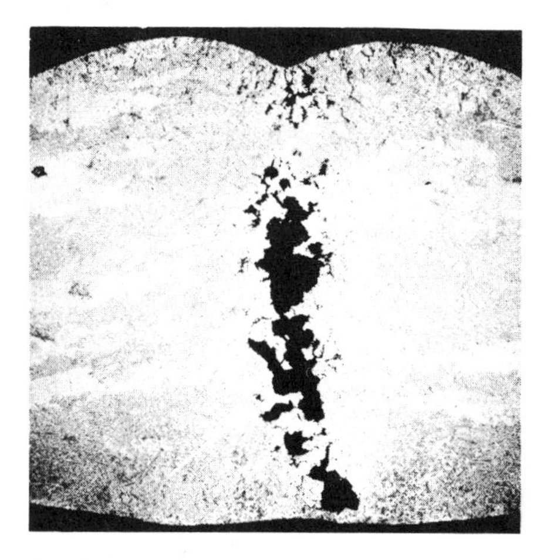

B CROSS SECTION OF (A) SHOWING EXTENT OF POROSITY

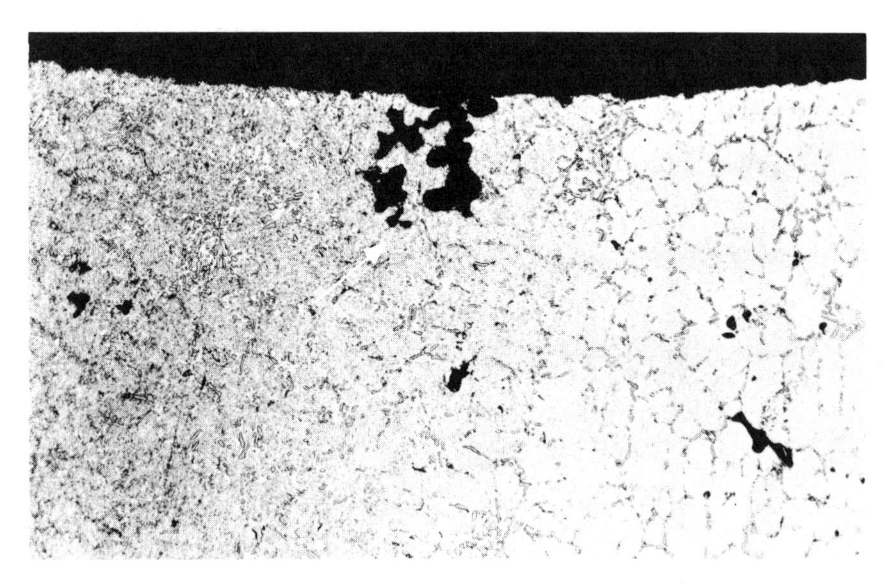

C MICROGRAPH OF CROSS SECTION SHOWING TYPICAL SHRINKAGE POROSITY

Figure A-25. Gas Porosity Discontinuity

- Metallurgical Analysis

Porosity in welds is caused by gas entrapment in the molten metal, too much moisture on the base or filler metal, or improper cleaning or preheating.

- NDT Methods Application and Limitations

  - Radiography Testing Method

    - Radiography is the most universally used NDT method for the detection of gas porosity in weldments.

    - The radiographic image of a "round" porosity will appear as oval-shaped spots with smooth edges, while "elongated" porosity will appear as oval-shaped spots with the major axis sometimes several times longer than the minor axis.

    - Foreign material such as loose scale, flux, or splatter will affect validity of test results.

  - Ultrasonic Testing Method

    - Ultrasonic testing equipment is highly sensitive and is capable of detecting microseparations. Established standards should be used if valid test results are to be obtained.

    - Surface finish and grain size will affect the validity of the test results.

- Eddy Current Testing Method

  - Normally confined to thin-walled welded pipe and tube.

  - Penetration restricts testing to a depth of more than 0.25 inch (6.35 mm).

- Liquid Penetrant Testing Method

  - Normally confined to in-process control of ferrous and nonferrous weldments.

  - Liquid penetrant testing, like magnetic particle testing, is restricted to surface evaluation.

  - Extreme caution must be exercised to prevent any cleaning material, magnetic (iron oxide), and liquid penetrant materials from becoming entrapped and contaminating the rewelding operation.

- Magnetic Particle Testing Method

  - Not normally used to detect gas porosity. Only surface porosity would be evident. Near surface porosity would not be clearly defined since indications are neither strong nor pronounced.

Unfused Porosity

- Category - Processing

- Material - Aluminum

- Discontinuity Characteristics

Internal. Wafer-thin fissures aligned parallel with the grain flow. Found in wrought aluminum that has been rolled, forged or extruded (Figure A-26).

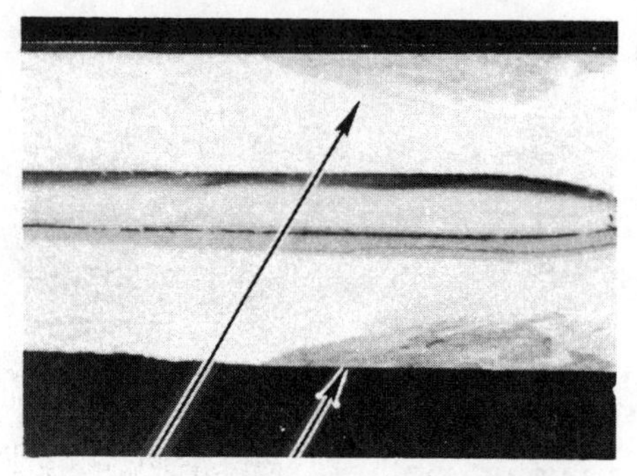

A. FRACTURED SPECIMEN SHOWING UNFUSED POROSITY

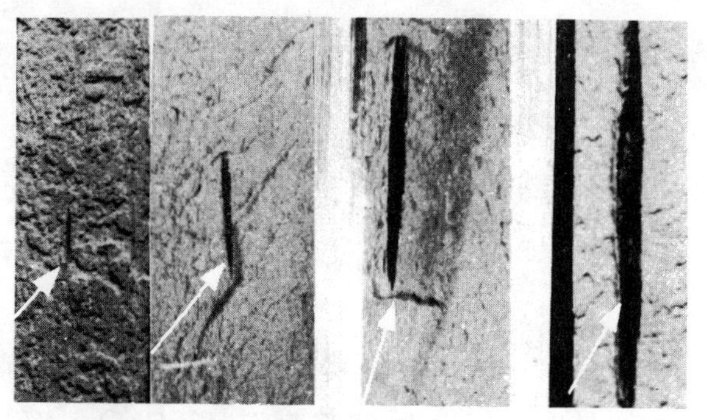

B. UNFUSED POROSITY EQUIVALENT TO 1/64 IN. (0.40 mm), 3/64 IN. (1.17 mm) 5/64 IN. (1.98 mm) AND 8/64 IN. (3.18 mm) (left to right)

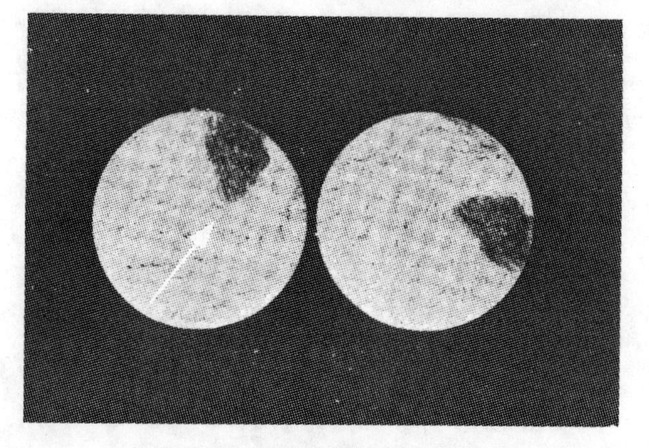

C. TYPICAL UNFUSED POROSITY

Figure A-26. Unfused Porosity Discontinuity

- Metallurgical Analysis

Unfused porosity is attributed to porosity in the cast ingot. During the rolling, forging or extruding operations it is flattened into a wafer-thin shape. If the internal surface of these discontinuities is oxidized or is composed of a foreign material, they will not fuse during the subsequent processing. This results in an extremely thin interface or void.

- NDT Methods Application and Limitations

  - Ultrasonic Testing Method

    - Used extensively for the detection of unfused porosity.

    - Raw materials may be tested in the "as-received" configuration.

    - Ultrasonic testing fixes the location of the void in all three directions.

    - Where the general direction of the discontinuity is unknown, it may be necessary to test from several directions.

    - Method of manufacture and subsequent article configuration will determine the orientation of the unfused porosity to the material surface.

  - Liquid Penetrant Testing Method

    - Normally used on nonferrous machined articles.

    - Unfused porosity will appear as a straight line of varying lengths running parallel with the grain. Liquid penetrant testing is restricted to surface evaluation.

- Surface preparations such as vapor blasting, honing, grinding, or sanding may obliterate possible indications by masking the surface discontinuities and thereby restricting the reliability of liquid penetrant testing.

- Excessive agitation of penetrant materials may produce foaming.

– Eddy Current Testing Method

  - Not normally used for detecting unfused porosity.

– Radiographic Testing Method

  - Not normally used for detecting unfused porosity. Wafer-thin discontinuities are difficult to detect by a method that measures density or that requires that the discontinuity be perpendicular to the X-ray beam.

– Magnetic Particle Testing Method

  - Not applicable. Material is nonferrous.

Stress Corrosion

- Category - Service

- Material - Ferrous and Nonferrous

- Discontinuity Characteristics

- Surface. Range from shallow to very deep, and usually follow the grain flow of the material; however, transverse cracks are also possible (Figure A-27).

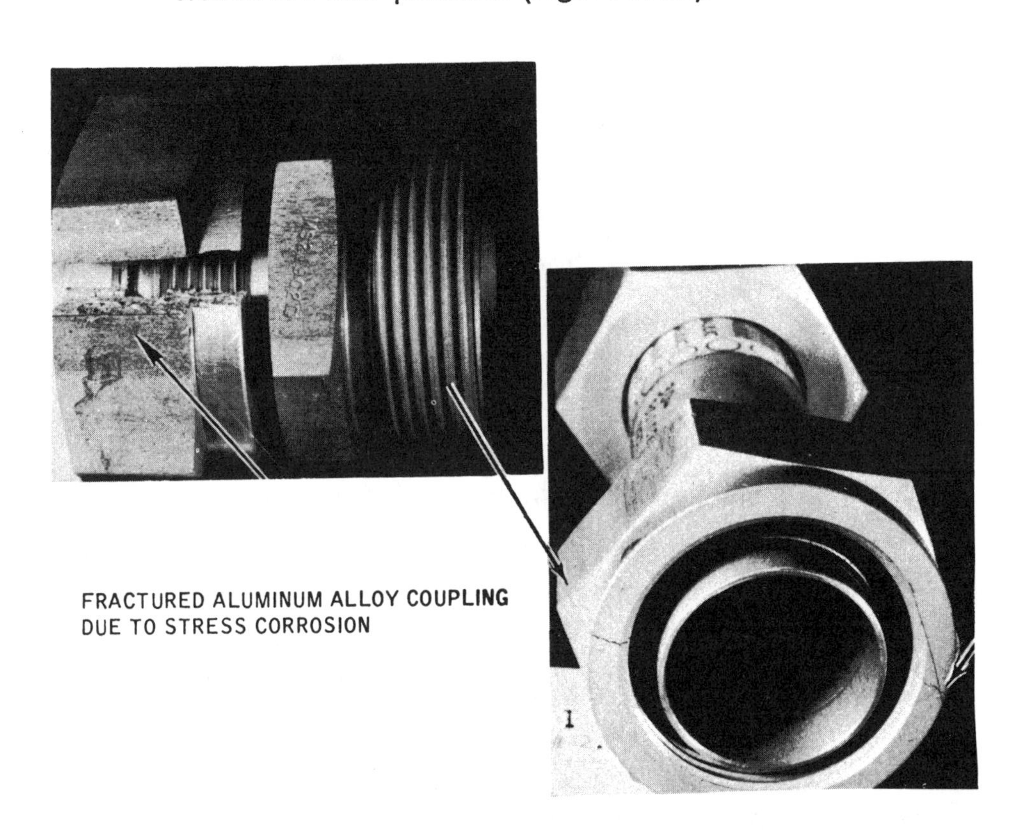

Figure A-27. Stress Corrosion Discontinuity

- Metallurgical Analysis

The following three factors are necessary for the phenomenon of stress corrosion to occur: 1) a sustained static tensile stress, 2) the presence of a corrosive environment, and 3) the use of a material that is susceptible to this type of failure. Stress corrosion is much more likely to occur at high levels of stress than at low levels of stress. The type of stresses include residual (internal) as well as those from external (applied) loading.

- NDT Methods Application and Limitations

  - Liquid Penetrant Testing Method

    - Liquid penetrant is normally used for the detection of stress corrosion.

    - In the preparation, application, and final cleaning of articles, extreme care must be exercised to prevent overspraying and contamination of the surrounding articles.

    - Chemical cleaning immediately before the application of liquid penetrant may seriously affect the test results if the solvents are not given time to evaporate.

    - Service articles may contain moisture within the discontinuity which will dilute, contaminate, and invalidate results if the moisture is not removed.

  - Ultrasonic Testing Method

    - Advanced techniques have been successfully used to detect stress corrosion and stress corrosion cracking in the nuclear industry.

    - Indications appear in a variety of amplitudes, shapes, and characteristics.

    - Interpretation is often difficult, requiring highly-trained operators.

- Eddy Current Testing Method

    - Eddy current equipment is capable of resolving stress corrosion where article configuration is compatible with equipment limitations.

- Magnetic Particle Testing Method

    - Not normally used to detect stress corrosion. Configuration of article and usual nonferromagnetic condition exclude magnetic particle testing.

- Radiographic Testing Method

    - Not normally used to detect stress corrosion. Surface indications are best detected by NDT method designed for such applications; however, radiography can and has shown stress corrosion with the use of the proper technique.

Hydraulic Tubing

- Category - Processing and Service

- Material - Aluminum

- Discontinuity Characteristics

  Surface and internal. Range in size from short to long, shallow to very tight and deep. Usually they will be found in the direction of the grain flow with the exception of stress corrosion which has no direction (Figure A-28).

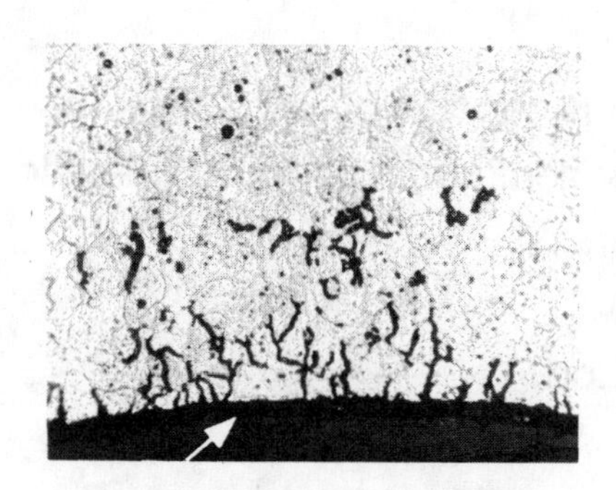

A INTERGRANULAR CORROSION

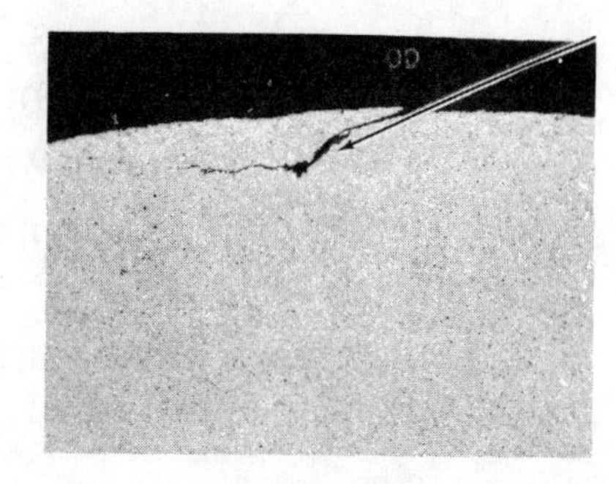

B LAP IN OUTER SURFACE OF TUBING

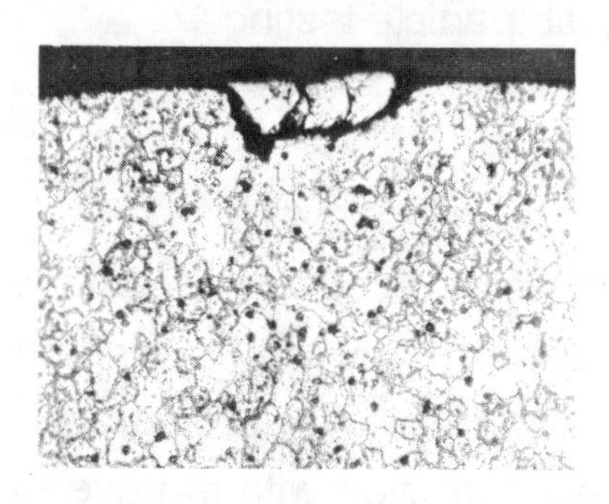

C EMBEDDED FOREIGN MATERIAL

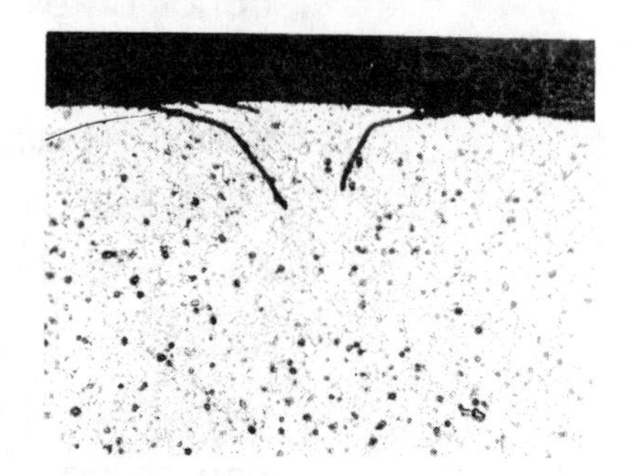

D TWIN LAPS IN OUTER SURFACE OF TUBING

Figure A-28. Hydraulic Tubing Discontinuities

- Metallurgical Analysis

  Hydraulic tubing discontinuities are usually one of the following.

  - Foreign material coming in contact with the tube material and being embedded into the surface of the tube.

  - Laps which are the result of material being folded over and not fused.

- Seams which originate from blowholes, cracks, splits, and tears introduced in the earlier processing, and then are elongated during rolling.

- Intergranular corrosion which is due to the presence of a corrosive environment.

- NDT Methods Application and Limitations

  - Eddy Current Testing Method

    - Universally used for testing of nonferrous tubing.

    - Heavier-walled tubing (0.25 inch or 6.35 mm and over) may not be successfully tested due to the penetration ability of the equipment.

    - The specific nature of various discontinuities may not be clearly defined.

    - Test results will not be valid unless controlled by known standards.

    - Testing of ferromagnetic material may be difficult.

    - All material should be free of any foreign material that would invalidate the test results.

  - Liquid Penetrant Testing Method

    - Not normally used for detecting tubing discontinuities. Eddy current is more economical, faster, and, with established standards, is more reliable.

- Ultrasonic Testing Method

  - Not normally used for detecting tubing discontinuities. Eddy current is recommended over ultrasonic testing since it is faster and more economical for this range of surface discontinuity and nonferrous material.

- Radiographic Testing Method

  - Not normally used for detecting tubing discontinuities. The size and type of discontinuity and the configuration of the article limit the use of radiography for screening of material for this group of discontinuities.

- Magnetic Particle Testing Method

  - Not applicable. Material is nonferrous.

Mandrel Drag

- Category - Processing

- Material - Nonferrous, Thick-walled Seamless Tubing

- Discontinuity Characteristics

  Internal surface of thick-walled tubing. Range from shallow, even gouges to ragged tears. Often a slug of the material will be embedded within the gouged area (Figure A-29).

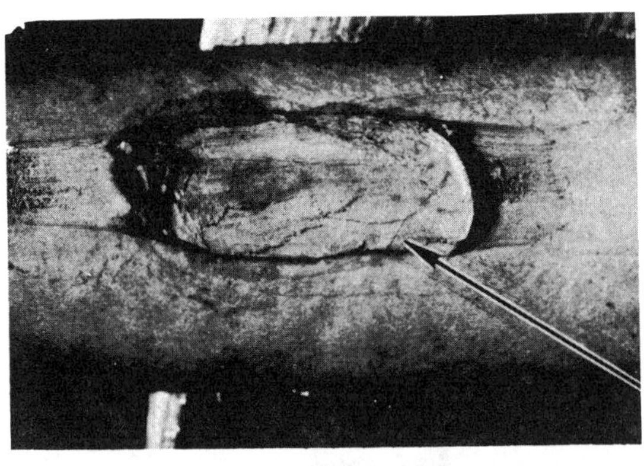

A. EMBEDDED SLUG SHOWING DEEP GOUGE MARKS

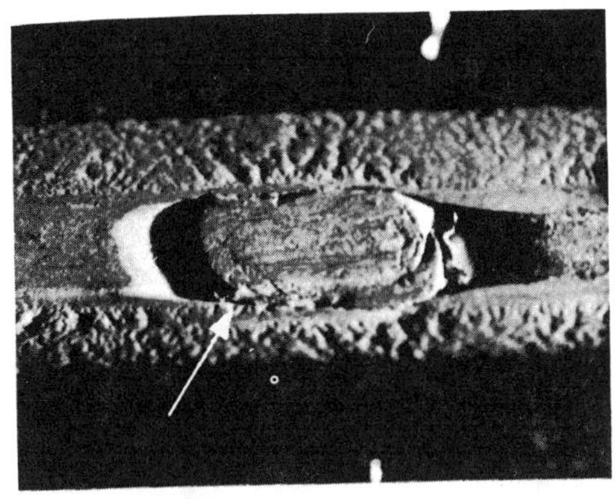

B. SLUG BROKEN LOOSE FROM TUBING WALL

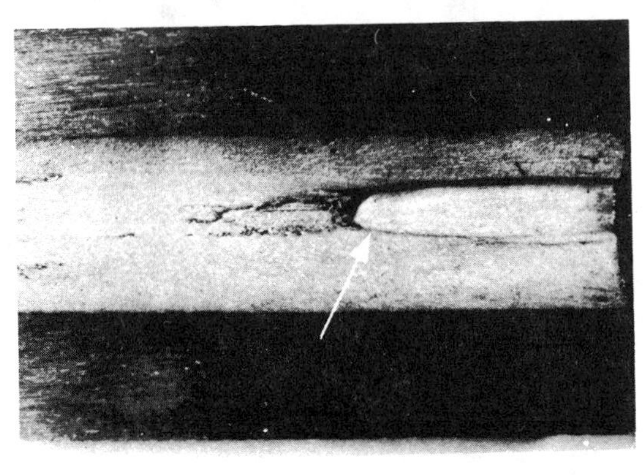

C. ANOTHER TYPE OF EMBEDDED SLUG

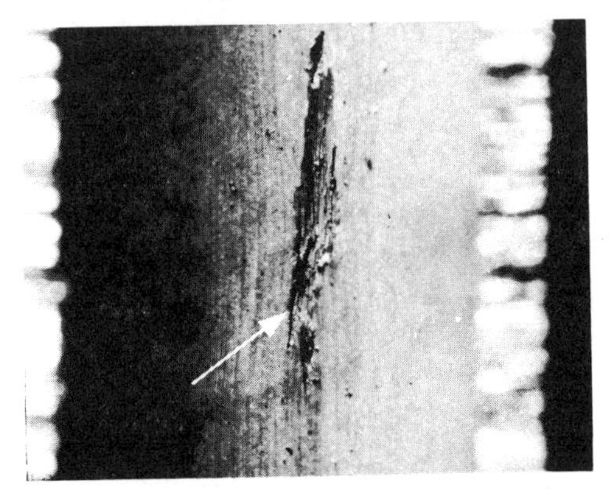

D. GOUGE ON INNER SURFACE OF PIPE

Figure A-29. Mandrel Drag Discontinuities

- Metallurgical Analysis

  During the manufacture of thick-walled seamless tubing, the billet is ruptured as it passes through the offset rolls. As the piercing mandrel follows this fracture, a portion of the material may break loose and be forced over the mandrel. As it does, the surface of the tubing may be scored or have the slug embedded into the wall. Certain types of material are more prone to this type of failure than others.

- NDT Methods Application and Limitations
  - Eddy Current Testing Method
    - Normally used for the testing of thin-walled pipe or tube.
    - Eddy current testing may be confined to nonferrous materials.
    - Discontinuities are qualitative indications and not quantitative indications.
    - Several factors simultaneously affect output indications.
  - Ultrasonic Testing Method
    - Normally used for the screening of thick-walled pipe or tube for mandrel drag.
    - Can be used to test both ferrous and nonferrous pipe or tube.
    - May be used in support of production line, since it is adaptable for automatic instrumentation.
    - Configuration of mandrel drag or tear will produce very sharp and noticeable indications on the scope.
  - Radiographic Testing Method
    - Not normally used although it has been instrumental

in the detection of mandrel drag during examination of adjacent welds. Complete coverage requires several exposures around the circumference of the tube. This method is not designed for production support since it is very slow and costly for large volumes of pipe or tube. Radiograph will disclose only two dimensions and not the third.

- Liquid Penetrant Testing Method

  - Not recommended for detecting mandrel drag since discontinuity is internal and would not be detectable.

- Magnetic Particle Testing Method

  - Not recommended for detecting mandrel drag. Discontinuities are not close enough to the surface to be detectable by magnetic particles. Most mandrel drag will occur in seamless stainless steel.

Semiconductors

- Category - Processing and Service

- Material - Hardware

- Discontinuity Characteristics

  Internal. Appear in many sizes and shapes and various degrees of density. They may be misformed, misaligned, damaged, or may have broken internal hardware. Found in transistors, diodes, resistors, and capacitors (Figure A-30).

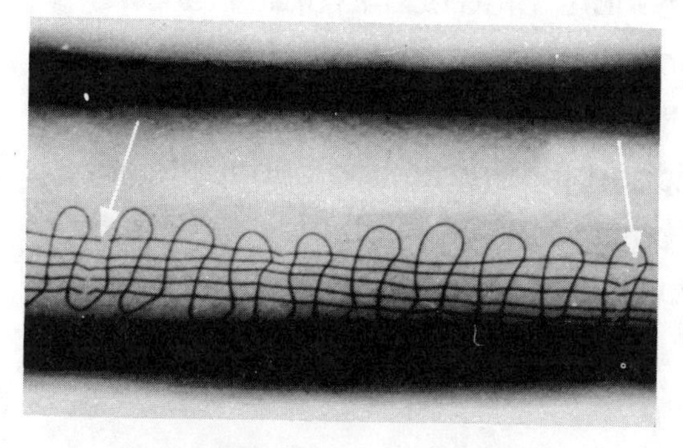

A. STRANDS BROKEN IN HEATER BLANKET

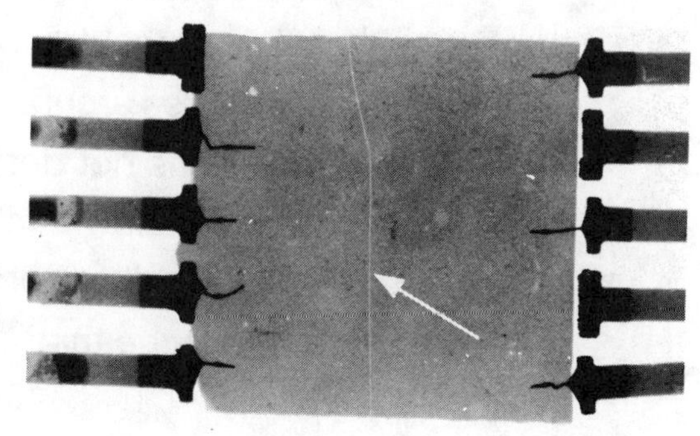

B. FINE CRACK IN PLASTIC CASING MATERIAL

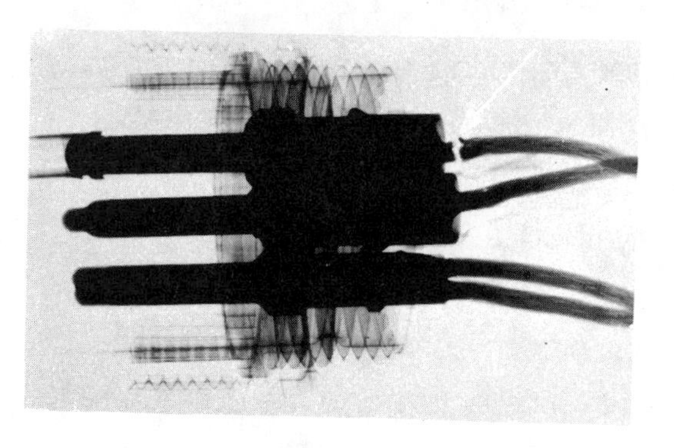

C. BROKEN ELECTRICAL CABLE

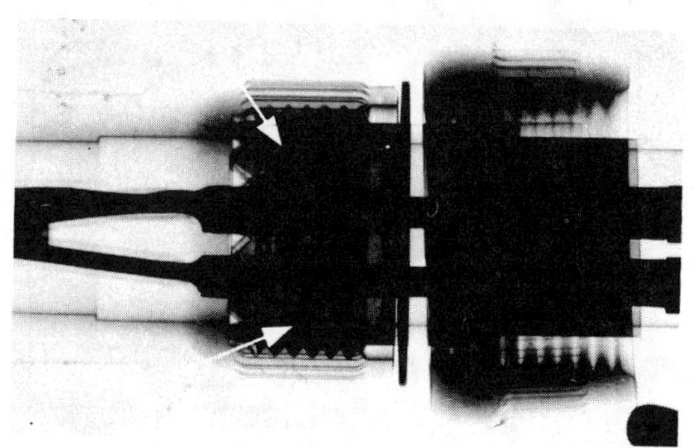

D. FOREIGN MATERIAL WITHIN SEMICONDUCTOR

Figure A-30. Semiconductor Discontinuities

- Metallurgical Analysis

  Semiconductor discontinuities such as loose wire, weld splash, flakes, solder balls, loose leads, inadequate clearance between internal elements and case and inclusions or voids in seals or around lead connections are the product of processing errors.

- NDT Methods Application and Limitations

  - Radiographic Testing Method

- Universally used as the NDT method for the detection of discontinuities in semiconductors.

- The configuration and internal structure of the various semiconductors limit the NDT method of radiography.

- Semiconductors that have copper heat sinks may require more than one technique due to the density of the copper.

- Internal wires in semiconductors are very fine and may be constructed from materials of different density such as copper, silver, gold, and aluminum. If the latter is used with the others, special techniques may be needed to resolve test reliability.

- Microparticles may require the highest sensitivity to resolve.

- The complexity of the internal structure of semi-conductors may require additional views to exclude the possibility of nondetection of discontinuities due to masking by hardware.

- Positive positioning of each semiconductor will prevent invalid interpretation.

- Source angle should give minimum distortion.

- Preliminary examination of semiconductors may be accomplished using a vidicon system that would allow visual observation during 360° rotation of the article.

- Eddy Current Testing Method

  - Not recommended for detecting semiconductor discontinuities. Nature of discontinuity and method of construction of the article do not lend themselves to this form of NDT.

- Magnetic Particle Testing Method

  - Not recommended for detecting semiconductor discontinuities.

- Liquid Penetrant Testing Method

  - Not recommended for detecting semiconductor discontinuities.

- Ultrasonic Testing Method

  - Not recommended for detecting semiconductor discontinuities.

Hot Tears

- Category - Inherent

- Material - Ferrous Castings

- Discontinuity Characteristics

  Internal or near surface. Appear as ragged line of variable width and numerous branches. Occur individually or in groups (Figure A-31).

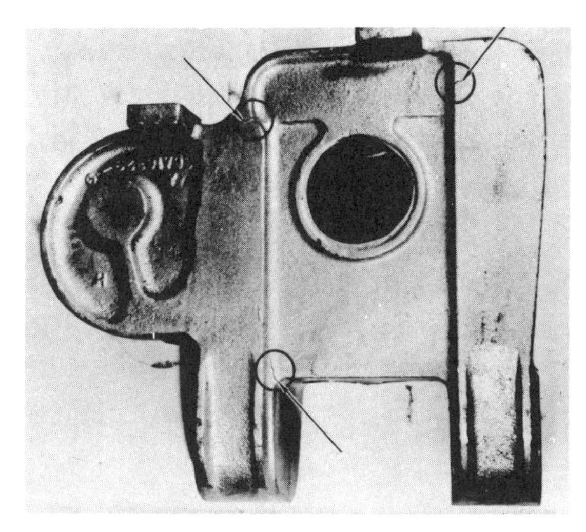

A. TYPICAL HOT TEARS IN CASTING

B. HOT TEARS IN FILLET OF CASTING

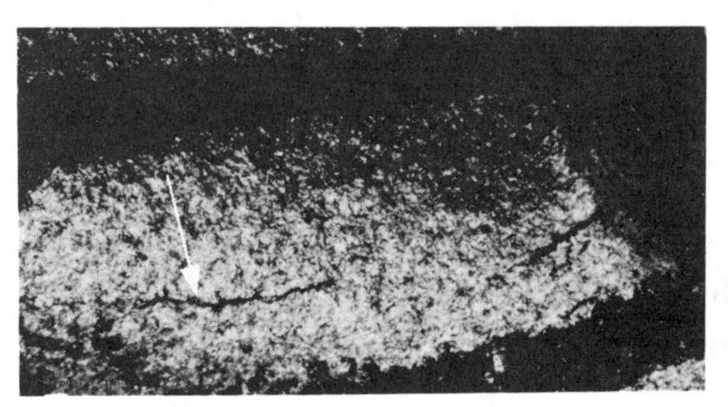

C. CLOSE-UP OF HOT TEARS IN (A)

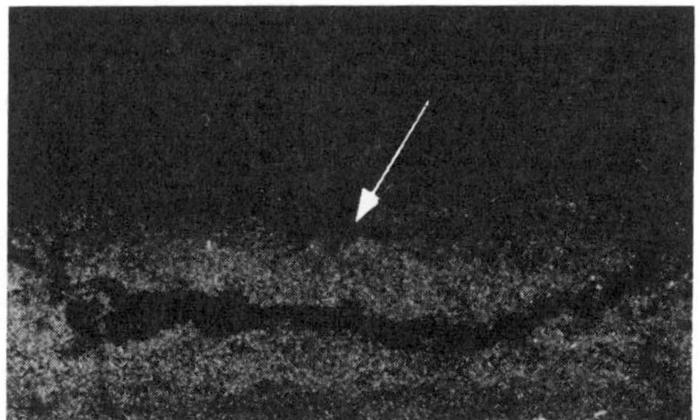

D. CLOSE-UP OF HOT TEARS IN (B)

Figure A-31. Hot Tear Discontinuities

- Metallurgical Analysis

Hot cracks (tears) are caused by nonuniform cooling resulting in stresses which rupture the surface of the metal while its temperature is still in the brittle range. Tears may originate where stresses are set up by the more rapid cooling of thin sections that adjoin heavier masses of metal which are slower to cool.

- NDT Methods Application and Limitations

  - Radiographic Testing Method

    - Radiographic testing is the first choice since the material is cast structure and the discontinuities may be internal and surface.

    - Orientation of the hot tear in relation to the source may influence the test results.

    - The sensitivity level may not be sufficient to detect fine surface hot tears.

  - Magnetic Particle Testing Method

    - Hot tears that are exposed to the surface can be screened with magnetic particle method.

    - Article configuration and metallurgical composition may make demagnetization difficult.

    - Although magnetic particle testing can detect near surface hot tears, radiography should be used for final analysis.

    - Foreign material not removed prior to testing will cause an invalid test.

- Liquid Penetrant Testing Method

  - Liquid penetrant testing is recommended for nonferrous cast material.

  - Method is confined to surface evaluation.

  - The use of penetrants on castings may act as a contaminant by saturating the porous structure and thereby affecting the ability to apply surface finish.

  - Repeatability of indications may be poor.

- Ultrasonic Testing Method

  - Not recommended for detecting hot tears. Discontinuities of this type, when associated with cast structure, do not lend themselves to ultrasonic testing.

- Eddy Current Testing Method

  - Capable of detecting surface hot tears. Metallurgical structure, along with the complex configurations, may require specialized probes and techniques.

Intergranular Corrosion

- Category - Service

- Material - Nonferrous

- Discontinuity Characteristics

Surface or internal. A series of small micro-openings with no definite pattern. May appear individually or in groups. The insidious nature of intergranular corrosion results from the fact that very little corrosion or corrosion product is visible on the surface. Intergranular corrosion may extend in any direction following the grain boundaries of the material (Figure A-32).

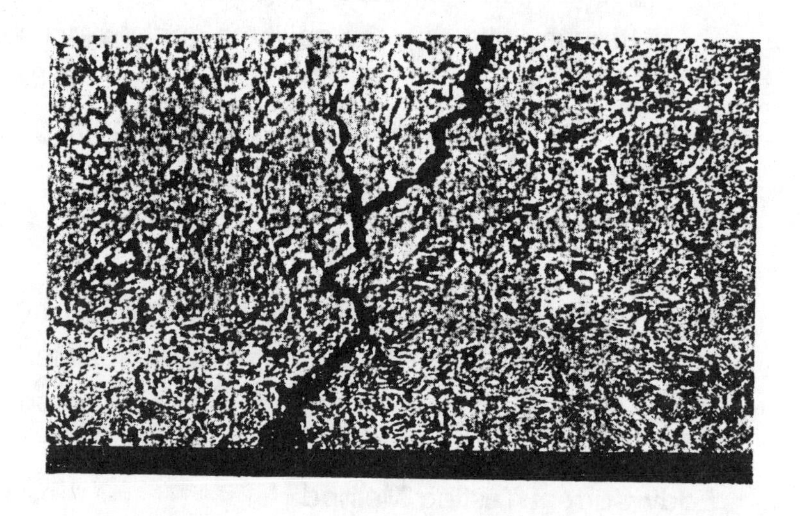

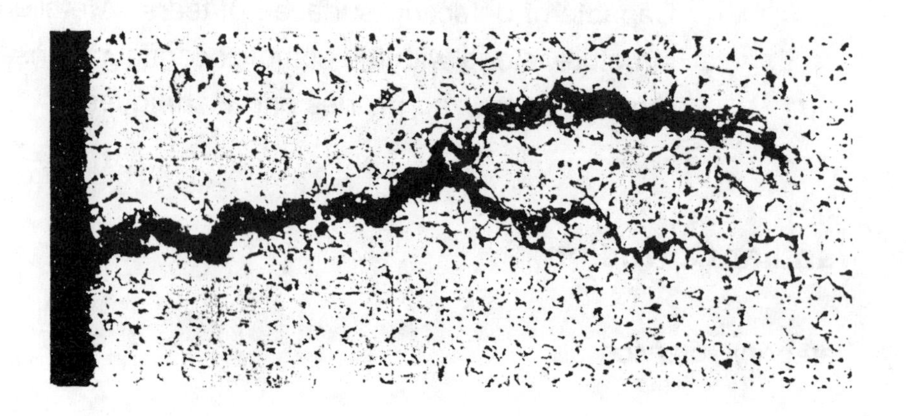

Figure A-32. Intergranular Corrosion Discontinuities

- Metallurgical Analysis

Two factors that contribute to intergranular corrosion are:

- Metallurgical structure of the material that is prone to intergranular corrosion, such as unstabilized 300 series stainless steel.

- Improper stress relieving or heat treat may create the susceptibility to intergranular corrosion.

Either of these conditions, coupled with a corrosive atmosphere, will result in intergranular attack.

- NDT Methods Application and Limitations

  - Liquid Penetrant Testing Method

    - Liquid Penetrant testing is the first choice due to the size and location of this type of discontinuity.

    - Chemical cleaning operations immediately before the application of liquid penetrant may contaminate the article and seriously affect test results.

    - Cleaning with solvents may release chlorine and accelerate intergranular corrosion.

    - Trapped penetrant solution may present a cleaning or removal problem.

- Ultrasonic Testing Method

    - Advanced techniques have been successfully used to detect stress corrosion and stress corrosion cracking in the nuclear industry.

    - Indications appear in a variety of amplitudes, shapes, and characteristics lending difficult interpretation.

- Eddy Current Testing Method

    - Eddy current can be used for the screening of intergranular corrosion.

    - Tube or pipe lend themselves readily to this method of NDT testing.

    - Metallurgical structure of the material may seriously affect the output indications.

- Radiographic Testing Method

    - Intergranular corrosion in the more advanced stages has been detected with radiography.

    - Sensitivity levels may prevent the detection of fine intergranular corrosion.

    - Radiography may not indicate the surface on which the intergranular corrosion occurs.

- Magnetic Particle Testing Method

    - Not recommended for detecting intergranular corrosion. Type of discontinuity and material restrict the use of magnetic particles.

APPENDIX B

GLOSSARY

Absolute Coil A test arrangement which tests the specimen without any comparison to either another portion of the test specimen or to a known reference.

Alternating A voltage, current or magnetic field that reverses direction at regularly recurring intervals.

Bobbin Coil A coil or coil assembly used for eddy current testing by insertion into the test piece; e.g., an inside probe for tubing. Also referred to as **Inside Coil** or **ID Coil**.

Coil Conductor wound in one or more loops to produce an axial magnetic field when current is passed through it.

Coil Spacing The axial distance between two encircling coils of a differential system.

Conductivity The willingness of a test circuit or test specimen to conduct current.

Coupling A measure of the degree to which the magnetic field of the coil passes through the test specimen and is affected by the magnetic field created by the flow of eddy currents.

Defect A discontinuity whose size, shape, orientation, location or properties limit, or have the potential to limit, the service of the test part in which it occurs or which exceeds the accept/reject criteria for a given design.

Defect Resolution A property of a test system which enables the separation of signals due to defects in the test specimen that are located in close proximity to each other.

Diamagnetic A material having a permeability less than that of a vacuum.

Differential Coil A test arrangement which tests the specimen by comparing the portion being tested with either another portion of the same specimen or to a known reference specimen.

Discontinuity Any interruption in the normal physical structure or configuration of a part, such as cracks, pits, dent, bulges, and wastage. A discontinuity may or may not affect the usefulness of a part.

Discontinuity, Artificial Reference discontinuities, such as holes, grooves, or notches, which are introduced into a reference standard to provide accurately reproducible sensitivity levels for electromagnetic test equipment.

Double Coil A test arrangement where the alternating current is supplied through one coil while the change in material condition is measured from a second coil.

Eddy Current A circulating electrical current induced in a conductive material by an alternating magnetic field.

Eddy Current Testing A nondestructive testing method in which eddy current flow is induced in the test object. Changes in the flow caused by variations in the specimen are reflected into a nearby coil, coils or other magnetic flux sensors for subsequent analysis by suitable instrumentation and techniques.

Edge or End Effect The disturbance of the magnetic field and eddy currents due to the proximity of an abrupt change in geometry (edge, end). The effect generally results in the masking of discontinuities within the affected region.

Effective Depth of Penetration The depth in a material beyond which a test system can no longer detect a change in material properties.

Effective Permeability A hypothetical quantity which is used to describe the magnetic field distribution within a cylindrical conductor in an encircling coil. The field strength of the applied magnetic field is assumed to be uniform over the entire cross section of the test specimen with the effective permeability, which is characterized by the conductivity and diameter of the test specimen and test frequency, assuming values between zero and one, such that its associated amplitude is always less than one within the specimen.

Electromagnetic Induction The process by which a varying or alternating current (eddy current) is induced into an electrically conductive test object by a varying electromagnetic field.

Electromagnetic Testing That nondestructive test method for engineering materials, including magnetic materials, which uses electromagnetic energy having frequencies less than those of visible light to yield information regarding the quality of the tested material.

Encircling Coil A coil, coils, or coil assembly that surrounds the part to be tested. Coils of this type are also referred to as circumferential, OD or feed-through coils.

External Reference Differential A differential test arrangement that compares a portion of the test specimen to a known reference standard.

Ferromagnetic A material which, in general, exhibits hysteresis phenomena, and whose permeability is dependent on the magnetizing force.

Fill Factor For an inside coil, it is the ratio of the outside diameter of the coil squared to the inside diameter of the specimen squared. For an encircling coil, it is the ratio of the outside diameter of the specimen squared to the inside diameter of the coil squared.

Flux Density A measure of the strength of a magnetic field expressed as a number of flux lines passing through a given area.

Henry The unit of inductance. More precisely, a circuit in which an electromotive force of one volt is induced when the current is changing at a rate of one ampere per second will have an inductance of one Henry. (Symbol: H)

Hertz The unit of frequency (one cycle per second). (Symbol: Hz)

High Pass Filter An electronic circuit designed to block signals of low frequency while passing high frequency signals.

IACS The International Annealed Copper Standard. A value of conductivity established as a standard against which other conductivity values are referred to in percent IACS.

Impedance The opposition to current flow in a test circuit or a coil due to the resistance of that circuit or coil, plus the electrical properties of the coil as affected by the coil's magnetic field.

Impedance Analysis An analytical method which consists of correlating changes in the amplitude, phase, or quadrature components (or all of these) of a complex test signal voltage to the electromagnetic conditions within the specimen.

Impedance-plane Diagram A graphical representation of the locus of points indicating the variations in the impedance of a test coil as a function of basic test parameters.

Inductance The inertial element of the electric circuit. An inductor resists any sudden change in the current flowing through it.

Inductive Reactance The opposition to current flow in a test circuit or coil when an alternating voltage source is applied and due solely to the electrical properties of the coil as affected by the magnetic field.

Inertia The property of matter which manifests itself as a resistance to any change in the momentum of a body.

Lift-off The distance between a surface probe coil and the specimen.

Lift-off Effect The effect observed due to a change in magnetic coupling between a test specimen and a probe coil whenever the distance between them is varied.

Low Pass Filter An electronic circuit designed to block signals of high frequency while passing low frequency signals.

Magnetic Field A condition of space near a magnet or current-carrying wire in which forces can be detected.

Magnetic Flux Lines A closed curve in a magnetic field through points having equal magnetic force and direction.

NDT (NonDestructive Testing) A testing method whereby the article or specimen being examined is undamaged by the process.

Noise Any undesired signal that tends to interfere with the normal reception or processing of a desired signal. In flaw detection, undesired response to dimensional and physical variables (other than flaws) in the test part is called "part noise."

Nonferromagnetic A material that is not magnetizable and hence, essentially not affected by magnetic fields. This would include paramagnetic materials having a magnetic permeability slightly greater than that of a vacuum and approximately independent of the magnetizing force and diamagnetic materials having a permeability less than that of a vacuum.

Paramagnetic A material having a permeability which is slightly greater than that of a vacuum, and which is approximately independent of the magnetizing force.

Permeability A measure of the ease with which the magnetic domains of a material align themselves with an externally applied magnetic field.

Permeability Variations Magnetic inhomogeneities of a material.

Phase Analysis An instrumentation technique which discriminates between variables in the test part by the different phase angle changes which these conditions produce in the test signal.

Phase Angle The angle measured in degrees that the current in the test circuit leads or lags the voltage. One complete cycle is equal to 360°.

Phase Shift A change in the phase relationship between two alternating quantities of the same frequency.

Probe Coil A small coil or coil assembly normally used for surface inspections.

Reference Standard A test specimen used as a basis for calibrating test equipment or as a comparison when evaluating test results.

Rejection Level The setting of the signal level above or below which all parts are rejectable or in an automatic system at which objectional parts will actuate the reject mechanism of the system.

Resistance The opposition to current flow in a test circuit or coil based on specific material properties and cross-sectional area and length of a conductor.

Response Amplitude The property of the test system whereby the amplitude of the detected signal is measured without regard to phase.

Saturation The degree of magnetization produced in a ferromagnetic material for which the incremental permeability has decreased substantially to unity.

Self-comparison Differential A differential test arrangement that compares two portions of the same test specimen.

Signal-to-noise Ratio The ratio of response or amplitude of signals of interest to the response or amplitude of signals containing no useful information.

Single Coil A test arrangement where the alternating current is supplied through the same coil from which the indication is taken.

Skin Effect A phenomenon where, at high frequencies, the eddy current flow is restricted to a thin layer of the test specimen close to the coil.

Standard A reference used as a basis for comparison or calibration; a concept that has been established by authority, custom, or agreement to serve as a model or rule in the measurement of quantity or the establishment of a practice or a procedure.

Standard Depth of Penetration The depth in a test specimen where the magnitude of eddy current flow is equal to 37 percent of the eddy current flow at the surface.